SORIN SONEA & MAURICE PANISSET
University of Montreal

A New Bacteriology

Jones and Bartlett Publishers, Inc.
Boston, Massachusetts and Portola Valley, California

Published in 1980 by Les Presses de l'Université de Montréal as introduction à la nouvelle bactériologie. This edition translated into English and revised under rights granted to Jones and Bartlett Publishers, Inc.

Editorial offices: 30 Granada Court, Portola Valley, California 94025.

Sales and Customer Service: 20 Park Plaza, Boston, Massachusetts 02116.

Library of Congress Cataloging in Publication Data

Sonea, Sorin.
  A new bacteriology.

Translation of: Introduction à la nouvelle
bactériologie.
  Bibliography: p.
  Includes index.
  1. Bacteriology. I. Panisset, Maurice. II. Title.
QR41.2.S6713 1983   589.9   83-13608

ISBN 0-86720-024-3 (Clothbound)
ISBN 0-86720-025-1 (Paperback)

STAFF FOR THIS BOOK: *Publisher:* Arthur C. Bartlett. *Designer and editor:* Elizabeth W. Thomson. *Illustrators:* Julia Gecha (figures) and Laszlo Meszoly (chapter openings). *Production:* Unicorn Production Services, Inc. *Composition:* Palatino, set by Achorn Graphic Services, Inc. *Printing and binding:* The Alpine Press, Inc.

Printed in the United States of America

9 8 7 6 5 4 3 2 1

Contents

Illustrations

Foreword

THIS IMPORTANT little book is indeed a manifesto for a new bacteriology. It will help stimulate what could be a long and enduring battle to improve our awareness of the role played by microbes in our environment and to change our traditional concepts about bacteria.

For three centuries bacteria have been considered to be alien and awe inspiring, even by sophisticated professors and dedicated students. Most of us still think that these tiny living beings are primarily germs and pathogens. They are often named by the symptoms they cause: the syphilitic spirochete, the plague bacterium, the cholera vibrio, and *Legionella*, the causative agent of Legionnaires' Disease. In this book, Dr. Sorin Sonea and his late colleague Dr. Maurice Panisset have begun to set the record straight. These organisms are not only our own ancestors but also are the basis of our life-support system. They supply our atmospheric gases, they cleanse our water supply, and, in general, they ensure us a livable environment.

Only recently have we recognized the true achievement and the communicative ability displayed by these all-pervading life forms. As Sonea and Panisset lucidly demonstrate, bacterial species are not to be conceived of in the same terms as those of plants and animals. A given visible bacterium is deficient, often lacking the genes it needs to do its job. It never functions as a single individual in nature. Therein lies the theme of this book:

vii

bacteria work in teams. They contain a *constant*, stable genetic system (the large replicon), but they function in the world by acquiring and exchanging a diverse set of *variable* genetic systems (several small replicons, including plasmids, viruses, and so forth). Small replicons are passed among bacteria with the rapidity and fluidity we associate with international telephone calls and transoceanic cables. We begin to comprehend not only the place of bacteria in the living world, but our error in considering them to be primitive because they are small. Their spores are comparable to satellites, their communications system forms a worldwide network.

Written in language that students and instructors can appreciate for its simplicity, clarity, and directness, this slim introduction to the vast and fascinating microbial world can also be enjoyed by those curious about the future of genetic engineering and biotechnology. The rich material on the history and philosophy of bacteriology should interest molecular biologists and geneticists who deal with bacteria in laboratories and hospitals. As we begin to understand the story told in this little book, we will do more than learn about pollution, water treatment processes, soil bacteria, and the inhabitants of cow rumen. In the end, we will find that *A New Bacteriology* is an original and profound statement about nature and about us.

Lynn Margulis

Preface

THE PRESENT BOOK, first published in French in 1980 with the title *Introduction à la nouvelle bactériologie*, is now presented in English. It has not only gone through translation, but also through a re-editing process that has been helpful in enriching the content of the original work.

I am very much indebted to Lynn Margulis, Professor of Biology at Boston University, whose mind-opening biological ideas were an inspiration during the work on this edition. Her encouragement and suggestions—for which I am very grateful—were essential to this book. She read the original French text and then all the successive manuscripts in English, giving excellent advice and positive criticism. She also helped with additional useful information, including many appropriate illustrations.

I also want to thank Elizabeth Thomson for editing that helped produce a smooth text from the initial English translation. It has been a pleasure to work with her on the final corrections and production of this book.

My thanks must also go to Arthur Bartlett, our publisher, for his encouragement and his patience with such a complicated re-edited translation.

My former co-author, Maurice Panisset, died in 1981, and so he could not enjoy the result of all this work. We both wanted to offer this book as a tool to all biologists and to new students in biology, biochemistry, and the health sciences, so that they would become familiar with the originality displayed by bacteria and would be aware of the decisive role of bacteria in the life processes on our planet.

Sorin Sonea

A NEW BACTERIOLOGY

Introduction

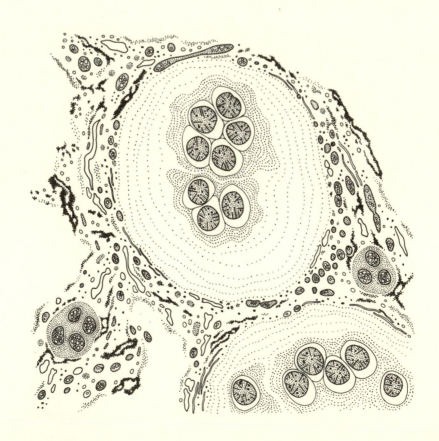

SEVERAL YEARS ago we advanced the hypothesis of a unitary bacterial world, an idea that forms the basis of a new bacteriology (47). This concept arose gradually over the course of twenty years of experimental and epidemiological work. It questions the familiar, traditional interpretations conveyed by current bacteriology textbooks, in which bacteria are described as members of numerous, distinct, and rather primitive species. Many authors limit themselves to a treatment of the molecular aspects of bacteria. Texts available today continue to include concepts that contradict recent discoveries (50). We believe that any contemporary study of the bacterial world must, from the very beginning, take into account its striking unique qualities that have been revealed over the course of the last thirty years.

Bacteria are prokaryotic microorganisms. In this book, the terms *prokaryote* and *bacterium* are used as synonyms. The structure and functions of bacteria distinguish them from all other living organisms, called *eukaryotes*. The principal differences between prokaryotes and eukaryotes are given in Table 1. These differences were pointed out as early as 1937 (14) but the fact that bacteria are distinguished from all other organisms was not generally accepted until more than twenty years later. After all, there was no long-standing oral tradition about bacteria as there was for animals and plants. They were described only after the development of the first microscopes, a little more than three centuries ago. They have been studied intensively only since the second half of the nineteenth century.

Bacteria are radically different from all other living creatures. Had they been discovered on Mars, their description would have

Table 1. Principal Differences Between Prokaryotes (Bacteria) and Eukaryotes

Characteristics	*Prokaryotes (Bacteria)* (Including eubacteria, cyanobacteria, archaebacteria, mycoplasmas, chlamydias, rickettsias)	*Eukaryotes* (Including animals, plants, fungi, and nucleated protists)
Cell structure		
Usual volume (in cubic μm)	Small: 0.01 to 30	Large: 1000 to 3,000,000
Grouping	Mostly unicellular, no tissues formed	Unicellular or multicellular, some with tissues
Heredity material	Free-floating circular DNA molecules, with no histones and no nuclear membrane	DNA attached to histones, with nuclear membrane
Cytoplasm	No intracellular motility	Intracellular motility
Membrane-bounded organelles	Absent	Mitochondria in most eukaryotes, chloroplasts in green algae and plants, other plastids in other algae; nucleus present
Cell walls	Present in most bacteria; contains polysaccharides and peptides	Of cellulose in plants, of chitin in fungi; chitin cellulose, protein in some protists, lacking in animals
Cytoplasmic membrane	Without sterols; no phagocytosis or pinocytosis	Contains sterols; phagocytosis and pinocytosis present

Flagella (related to motility)	Simple composition (one protein); move by rotation	9+2 (or 9+0 or 6+0) tubular flagella (undulipodia); more than 50 proteins; move by undulation
Morphological variety	Very limited	Very rich
Genetics		
Number of genes per cell	Small: 1000 to 5000; haploid and incomplete cells (for differentiation and evolution); very few repressed genes	Large: 200,000 to 3,000,000; diploid cells; generally only 4–6% genes actively producing protein at a time
Grouping of genes	Physical and functional separation of stable, essential genes (94–99%) in one large DNA molecule (called a "chromosome" by bacteriologists, "nucleoid" by electron microscopists, "genophore" by biologists, "large replicon" by geneticists), and of smaller DNA molecules of intracellular, visiting, unstable, nonessential genes (1–6% of the genome) as plasmids or in prophages (small replicons)	No DNA molecules specialized with transmissible favorable genes (exceptionally, viruses may play such a role)

Table 1. *Principal Differences Between Prokaryotes (Bacteria) and Eukaryotes (continued)*

Characteristics	Prokaryotes (Bacteria)	Eukaryotes
Gene pool	Each bacterium has access by undirectional transfer (with diminishing probabilities as metabolisms differ) to most or probably any bacterial gene in nature; gene exchanges are frequent, normal phenomena in bacteria outside the reproduction period and confer on them a strong solidarity. Usually a very limited number of genes may be exchanged for a generation of cells (fewer than 1%); in this way there is no sharp discontinuity between generations. There is no Mendelian sexuality; there were no dead direct ancestors of presently living bacteria (a common clone characteristic).	Gene exchanges (50% per generation and at random) in the usual Mendelian sexuality are limited to the members of the species involved and there is a complete discontinuity between generations; the ancestors of the living individuals are dead.
Metabolism		
Variety of types	Very large	Limited
Protein synthesis	Short-lived mRNA, smaller ribosomes tolerating more translation errors	Long-lived mRNA, larger ribosomes preventing some translation errors

Atmospheric nitrogen fixation	By a few different types of bacteria	Never performed
Photosynthesis	By different types of bacteria, from strict anaerobes to aerobes, with a variety of pigments	By chloroplasts (probable descendants of endosymbiotic cyanobacteria) present in plants and in a few other eukaryotes
Energy	Many different types of fermentation and respiration, the latter including the use of oxygen	Very limited fermentation; oxygen respiration
Oxygen tolerance	From extremely sensitive (toxicity) to tolerant and even dependent	Oxygen dependent

Functions

At the cell level Multiplication	Uninhibited successive divisions into two identical copies under favorable conditions	By mitosis and meiosis (many are sexual and need two parents)
Communicating genes	Each bacterial cell, independently of reproduction, has receptors for visiting genes and sends genes to other strains.	Gene exchange is regulated by Mendelian sexuality, related to reproduction in animals and plants (not in fungi and protists)
At the team level	Very frequent efficient associations of different bacteria, with labor division	With the exception of lichens, very few associations of different eukaryotes, with reciprocal advantage

Table 1. Principal Differences Between Prokaryotes (Bacteria) and Eukaryotes (continued)

Characteristics	Prokaryotes (Bacteria)	Eukaryotes
At the planetary level	Problem-solving capacities based on dispersion of genes and their choice, amplification, and direction by selective pressures. Life-supporting general activity	Exchange of genes during the reproduction process in animals and plants
Evolution		
Duration	Longer (started 3–3.5 billion years ago)	Shorter (started over 1 billion years ago)
Mechanism	All cells are germinative (a clonal way of life), and all bacteria might be considered to be a single clone. Strain variety developed inside this clone. There is little genetic isolation, as a huge gene pool seems to unify all bacteria. No extinction of specialized types, as genetic correction and reversible evolution are possible. There seems to be an evolution toward more gene exchanges, more communication and cooperation between bacteria. There	Most cells are not germinative once they differentiate. Species are isolated genetically, as they possess only their limited gene pool. Extensive extinction of past species occurs as evolution of eukaryotes is irreversible. There seems to be no guided evolution; each species acts and evolves rather anarchically.

is adaptation to all the physico—chemical characteristics of the planet and active evolutionary involvement that improves the life-supporting capacity of the general environment.

Differentiation

By acquisition of visiting genes from other types of bacteria (part of the planetary gene pool), with or without replacement of intracellular genes, or by loss of a plasmid or a prophage. There is no permanent repression of the majority of intracellular genes.

By repression of the majority of genes in a cell and the specific derepression of the genes necessary to that type of cell

7

been much more dramatic and the bizarre quality of their natural history, which often seems like science fiction, would not have been missed. Even today, bacteria are still described and discussed in nineteenth century terms.

Bacterial cells are usually much smaller than eukaryotic cells. Their structure is far less complex; they contain no nuclei, mitochondria (organelles essential for respiration in eukaryotes), or plastids (organelles essential for photosynthesis in plants and algae; in this book the term *algae* applies only to eukaryotes). Although some bacteria are multicellular, they do not associate to form tissues or organs; they are primarily unicellular beings. As isolated cells, they lack enough genetic material to undergo differentiation or evolution in the same way that eukaryotes do, and each bacterial cell has limited metabolic capacities. Compared to cells of animals or plants, bacterial cells are genetically incomplete. As compensation for this, each cell possesses mechanisms providing access to hereditary bacterial material from outside, and they easily form associations with other types of cells. It has become clear to us that bacteria, in addition to carrying out their individual and localized team activities, together form a planetary entity of communicating and cooperating microbes, an entity that, we think, is both genetically and functionally a true superorganism. It is at the social level—the level of their associations—that bacteria manifest their exceptional capacities and play a major role in nature (47,50,51).

Bacteria do not fuse to form new individuals. Each genetic exchange between different types of bacteria involves a small number of genes that are short sequences of deoxyribonucleic acid (DNA) containing hereditary information, usually for one protein molecule, mostly an enzyme, per gene. All bacteria reproduce asexually by division, but gene exchange takes place frequently in nature at any time and is not related to reproduction (8,10,13,23,28,31,42–44,57). Thus bacteria have a means of exchanging their intracellular genetic information only if necessary and with genes of their choice. They do not impose on one small cell the burden of gene accumulation. Bacteria communicate by passing genes from one type of cell to another, a behavior that differentiates prokaryotes from eukaryotes more than any other characteristic, just as speech distinguishes humans from all animals. Thus each bacterial cell benefits from a very large gene pool (all the genes of a population that exchanges genes among its members) that probably extends to all bacterial genes on Earth. This gene pool may be compared to the central data bank of a large electronic communications network.

The evolution of bacteria has lasted much longer than the evolution of eukaryotes and shows many original aspects. Genetic corrections are possible in bacteria, and therefore evolution is not irreversible for them. The bacterial world, although dispersed, has remained united by gene exchanges and by its common origin. The lack of genetic isolation in bacteria has meant that their evolution has not lead to genetically isolated species formation. In eukaryotes, a species consists of very similar beings, usually interbreeding and thus participating in the same gene pool, isolated genetically from all other beings. We see in bacteria a clonal way of life (where by a clone we mean an entity formed by cells all originating from a common ancestral cell), achieved by means of uninterrupted divisions each time into two identical cells. Probably all bacteria are descendants of the first cell on Earth; they are kept united by the possibility of sharing their genes.

Bacteria use the resources of the global bacterial gene pool as a central information source, and the extensive exchanges of genes that take place are amplified and directed by continual selection among the cells. Thus the bacterial world functions as a vast communications network to promote life in all possible niches of the planet by establishing associations of different types and genetic permutations whenever conditions are favorable. Bacteria thus seem to maintain the life-promoting stability of the chemical features surrounding the living world (30,47). Bacteria have adopted the surface of our planet as their giant protective shell and as a provider of water, nutrients, and gases. They transform dead cells into mineral substances that may be readily assimilated by plants, and they fix and directly use atmospheric nitrogen, making it available for eukaryotes. All existing photosynthesis evolved first in bacteria, including the photosynthesis performed today by plants and algae (36).

As a result of individual and collective evolution, gene exchanges, the interplay of complementary metabolisms, and long selective pressure, bacteria associated in local teams, pooling their metabolisms to compensate for the limited amount of hereditary information in each cell. They have progressively changed their immediate environment, the humid or liquid surfaces of our planet, and they have even altered its atmosphere, providing it with oxygen and probably with stability (30). The biosphere, the place where life exists, was thus already relatively favorably balanced about one and a half billion years ago when the first eukaryotic cells were formed, most likely through successive symbioses of different bacteria, as proposed by the serial endosymbiotic theory (32,35,55). Symbiosis is a lasting association, mutually advantageous under certain conditions, formed among different beings.

Today, with the minor exception of bacterial agents of food spoilage and of infectious diseases, the great majority of bacteria are essential for the well-being of the biosphere. Through their associations not only with other bacteria but also with all organisms, all bacteria on Earth constitute an essential, actively constructive factor in the biosphere. They contribute more than any living group to the fertility of our planet and its stable ecology. No plants or animals would survive even a few years without bacteria in nature, whereas bacteria have survived at least two billion years without plants or animals.

These aspects of the natural history of bacteria are not usually stressed in bacteriology textbooks. For many bacteriologists, bacteria are still the neglected Cinderella of biology. We therefore felt a need to introduce this new bacteriology to give just measure to bacterial natural history—its originality, its complexity and its all-pervading role. This book attempts to provide a coherent framework into which more specialized or advanced scientific information from other authors can easily be integrated, even when the general premises may well be different. We hope that this book will also help close the communications gap between the two main types of bacteriologists: those who study, utilize, or fight particular bacteria in nature or in industry, and those who probe the mysteries of their intracellular mechanisms through the intensive molecular study of a few types of bacteria that have become choice laboratory subjects. Because their respective concepts are so different, and because these two groups usually ignore each other, there is no generally accepted contemporary definition of bacteria, and an identity crisis exists in bacteriology

(53). We hope we can contribute to this area by building bridges between these polarized positions.

Our radical restructuring of such an important area in biology is bound to present some imperfections and weaknesses. We wished to stress the originality of the bacterial world, and have therefore given to this book a framework very different from that of previous texts. We hope to receive criticisms and suggestions from the readers of this first English edition of a pedagogical tool that, in fact, also constitutes a kind of manifesto for a new bacteriology.

CHAPTER 1

The History of Bacteriology

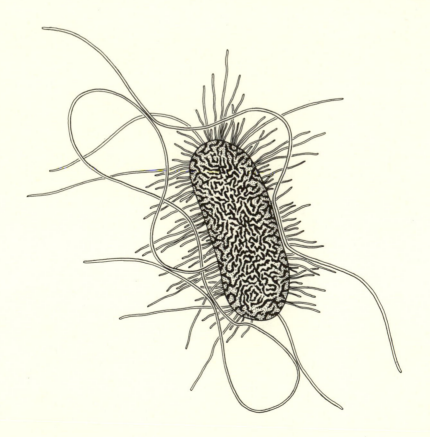

MODERN BACTERIOLOGY has suffered because bacteriologists are generally ignorant of its history. Knowledge of the evolution of a science will reveal the dynamics of its main orientations, its weaknesses, and its strengths. The history of bacteriology can be outlined in six stages. Each is characterized by its own particular technology, methodology, and accumulation of scientific knowledge.

Earliest Observations

For centuries, people had observed phenomena that we now know to be caused by bacteria. Over the course of time, a large body of empirical facts was accumulated. Bacterial phenomena such as fermentation, putrefaction, and infectious diseases in humans, animals, and plants, were selectively utilized or avoided for thousands of years before the discovery of bacteria. Such empirical observations and their applications formed the basis for the first scientific experiments that marked the beginning of bacteriology.

Three hundred years ago in Holland, the naturalist Anthony van Leeuwenhoek (1632–1723) constructed a microscope that enabled him to observe what he called "animalcules" (what we call *microbes*, organisms that can be seen only with the aid of a microscope). At last, bacteria were actually seen in samples of substances whose transformation had long been observed. However, it was only much later that the presence of van Leeuwenhoek's animalcules was considered to be the cause rather than the result of these transformations of matter. On the basis of his drawings

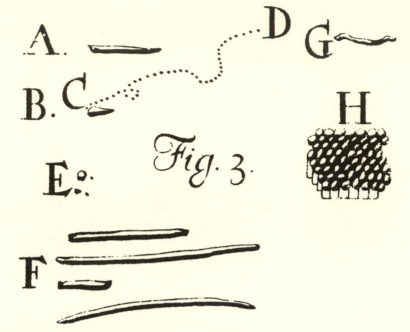

Figure 1–1. Van Leeuwenhoek's animalcules.

and observations, van Leeuwenhoek is recognized as the first to have seen bacteria. His drawings show the three basic shapes of the bacterial cell: spherical, rod-like, and spiroid (Figure 1–1). For nearly two centuries, bacteria remained as inaccessible for experimental study as the planets in the sky, whose observation was then also becoming possible as a result of parallel developments in the construction of telescopes.

First Experimental Studies

At the end of the eighteenth century, Lazzaro Spallanzani, naturalist and professor in northern Italy (1729–1799), and other scientists were able to obtain the multiplication of unidentified bacteria in organic liquids, such as broth or other extracts, that had previously been boiled. They observed a surprisingly rapid increase in the microbial population in one or two days. To this day, their simple method of artificial culture constitutes the basis of bacteriological study. A truly scientific method of bacterial culture was devised in Paris in the nineteenth century by Louis Pasteur (1822–1895), who was able to obtain so-called "pure bacterial cultures." He did this by carrying out serial dilutions of samples in a nutrient liquid that had been sterilized by boiling, and then retaining the cultures from the highest dilutions. The cultures from these highest dilutions had most probably only one bacterium left, and this bacterium then grew. Pasteur also showed that every living cell originates from another living cell, thereby eliminating the then popular concept of spontaneous generation, which had been based on insufficiently controlled observations.

> *Momentous discoveries of Pasteur: scientific basis of microbiology; microbial origin of infectious diseases*

Victory Over Infectious Diseases; the Concept of Bacterial Species

Pasteur was able to prove the bacterial origin of several infectious diseases in animals. In the years that followed, the methods he devised and the results of his experiments led to the discovery, isolation, and characterization of the majority of bacteria that cause diseases in humans, animals, or plants. These bacteria are called pathogens, from *pathos* (illness) and *gen* (causing). Bacteria were therefore seen as playing a predominantly nefarious role, causing food spoilage as well as infectious diseases, then the greatest single cause of mortality. Consequently, Pasteur and other researchers of his period, including his students and emulators, directed their attention towards the search for methods of prevention and treatment of bacterial infections. Many new or improved techniques were systematically recommended and applied. The isolation of the contagiously ill, long practiced on an empirical basis, was organized more efficiently. Chemicals toxic to most bacteria, called antiseptics, were discovered and used. The widespread use of vaccines (immunizing preparations injected to protect for years against a specific disease) and sera (the liquid part of the blood containing immediately available but short-lived antibodies, proteins acting specifically against infectious diseases) was highly successful. Much later the antibiotics, exclusively antibacterial drugs of biological origin, were added to this set of defenses. It was by means of these methods that a large portion of humanity was freed from serious infectious diseases, thus permitting a dramatic increase in human life expectancy and generating the population explosion that mankind is experiencing in the twentieth century.

Another unexpected consequence of practical success in the fight against bacterial disease was a premature and lasting decrease in scientific interest in bacteria for their own sake. Instead, scientists concentrated on the study of the pathogenic properties of bacteria and the mechanism of resistance to infection in humans and animals. Their focus turned to developing and refining methods and techniques of diagnosis and immunization, that is, to the field of medicine. The direction of bacteriology might have been quite different had Pasteur, after discovering pathogenic bacteria, extended his studies to the global role of bacteria.

> *The uniqueness of bacteria and a premature classification*

Pasteur, nevertheless, did recognize the uniqueness of bacteria. He concluded that it would be premature to determine organizational categories for the purpose of integrating bacteria into a classification scheme similar to the Linnean scheme already used for the plant and animal kingdoms. Other major discoverers of pathogenic bacteria were less realistic. They soon pigeon-holed bacteria into so-called "species" primarily on the basis of their respective practical roles or properties, such as the capacity to curdle milk or to infect animal or plant tissues.

> *Erroneous proliferation of categories of bacterial species and the controversy of pleomorphism versus monomorphism*

The isolation and the rapid laboratory multiplication of bacteria, even from single cells, convinced a growing number of investigators that they were dealing with distinct species, that is, with biological categories defined by similar properties and genetic isolation, whose apparent number has continued to grow. The practical stability of isolated and laboratory-grown strains of bacteria seemed to corroborate the concept of bacterial species exhibiting fixed, distinctive traits.

However, several bacteriologists had observed that the structure, form, and some biochemical reactions of certain bacterial strains (all members of a pure bacterial culture, supposed to have originated from one cell) could be modified by altering the growth medium. They permitted themselves to generalize on the basis of these observations, which had resulted in part from the use of deficient methods and techniques, and went so far as to propose a general theory of bacterial *pleomorphism* (*pleo*, more; *morph*, structure or form), implying that different types of bacteria were only different manifestations of a unified bacterial world. At the end of the nineteenth century, a bitter controversy arose between the "fixists" or monomorphists and the pleomorphists. The intransigence and verbal violence displayed by the various factions in this conflict resembled certain historic theological quarrels. It was proved irrefutably that some of the observed cases of pleomorphism were, in fact, due to contamination. At that time (1877–1890), it was not technically possible to demonstrate that other observations of pleomorphism might have been due to true modification phenomena peculiar to bacteria.

The theory of pleomorphism, which implied a certain unity in the bacterial world, was globally discredited, mostly by Sergei Winogradsky (1856–1953) who was working in Russia and France at the end of the nineteenth century. These circumstances explain the reserve and sometimes outright hostility that subsequently greeted all ideas that might point to the concept of a unified bacterial world. Conclusive new facts were required in order to question and overcome the scientific anathema surrounding any theory that even indirectly evoked pleomorphism. These new facts were generated only much later by advances in bacterial genetics.

The long reign of fixism

In the meantime, fixism (the theory that bacterial strains do not change in nature) and the concept of typical bacterial species have dominated bacteriological theory, delaying until now the development of a dynamic and unitary concept of the bacterial world. In the 1974 edition of *Bergey's Manual of Determinative Bacteriology*, the most widely-accepted book for the nomenclature, identification, and classification of bacteria (11), we can read (p. xvi), "For most groups of bacteria, genera and species are the only categories that can now be recognized and defined with reasonable precision . . . ."

Doubts about the concept of species in soil bacteria

Teams of soil bacteria

At the beginning of the twentieth century, Sergei Winogradsky in France and Martinus Willelm Beijerinck (1851–1923) in Holland, both pioneers in soil bacteriochemistry and geochemistry, demonstrated the existence of profound differences between soil bacteria and bacteria involved in infection. Soil bacteria exist as mixed groups carrying out complementary functions; many of their strains resist isolation on artificial culture media. Bacteria inhabit the soil in very high concentrations; they ensure its fertility and play a key role in nature by recycling carbon and nitrogen. For soil bacteriologists, the so-called species of a bacterium is often unclear and of no practical interest. Because of their complementary roles, soil bacteria can be considered as teams or cooperating entities. Soil microbiologists frequently are unable to isolate and characterize bacterial species. However, medical microbiologists who were dealing with the relatively limited group of parasitic bacteria imposed acceptance of the concept of stable species for all bacteria. Bacteria found in soil are, in fact, much more representative of the bacterial world as a whole than are the parasitic, infection-causing strains. Despite recognition of the ecological role of efficient, well-coordinated localized teams of bacteria containing different strains, soil microbiologists did not question the dogma of primitive, isolated species, and they did not realize that soil bacteriology was concerned with by far the greatest proportion of the global bacterial population. Had they recognized, from the beginning, the existence of a more complex level of organization among these bacteria and emphasized their decisive role as cooperating teams in the maintenance of the biosphere, it is probable that the whole concept of bacteriology could have changed at an early stage of the discipline.

Basic discoveries concerning bacteria; the paradoxical neglect of their study in general

Molecular Biology and Bacteria

The name "molecular biology" was given to the explanation, in chemical terms, of the genetic phenomena of replication, mutation, and DNA recombination. Reproduction and growth were described in terms of interactions between small food molecules and macromolecules such as proteins and nucleic acids, present in all cells. During the golden age of molecular biology, from 1950 to 1970, scientists uncovered the mechanisms of protein and nucleic acid synthesis. The universal genetic code was deciphered: rules that relate the sequence of nucleotides in nucleic acids to the sequence of amino acids in proteins. Continuous progress has been made towards an accurate specification of the relation of the sequences in bacterial DNA molecules to their function in the synthesis of proteins and all other cell constituents. The successive problems posed in molecular biology were approached, studied, and resolved in a spirit of enthusiasm characteristic of this exhilarating period in the history of biology. The fundamental unity of life was established. The most striking examples of this unity are a nearly universal genetic code, and widespread similarities of macromolecule synthesis and of several mechanisms of cell regulation and differentiation (25). Other discoveries unveiled specific bacterial phenomena that have led us to revise in depth our conception of the bacterial world (3,8,10,12,15,17, 18,20,21,23,28,31–44,49,53,57). However, we feel that, with a few rare exceptions, the scientific approach of molecular biology has been infused with an overly exclusive reductionism. It implied that most of the important problems of contemporary biology had physico–chemical solutions that were readily attainable through the use of modern technology. This optimistic perspective led physicists, chemists, and biochemists to join the biologists who were pioneering the field of molecular biology.

For an understanding of the bacterial world, all these new discoveries added decisive support to the earlier ideas of two important scientists. As early as 1937, E. Chatton, working in Strasbourg, had underlined the fundamental differences between the group of organisms comprising the more familiar bacteria and blue-green algae, which he named *prokaryotes*, and all other living organisms, which he designated as *eukaryotes* (14). Electron microscopy, which began on biological entities before 1950, entirely confirmed Chatton's assertion. Bacterial genetics equally proved to be very original. In 1928, Frederick Griffith, working in England, had surprised and slightly shocked microbiologists, particularly those working in the health sciences, when he proved that a type of pathogenic bacteria, the pneumococci, could aquire new hereditary properties from contact with dead pneumococci of another type. Unfortunately, he did not explore all the possibilities generated by this epoch-making discovery. Like his colleagues, he was a physician more interested in infections than in the bacteria that cause them.

Early perception of the possibility of unity in the bacterial world

From the outset, Griffith's discovery was interpreted by several medical microbiologists as the first demonstration that bacteria could exchange genes without recourse to the mechanisms of a two-parent, meiotic, sexual process. This phenomenon of *transformation* was used at the Rockefeller Institute in New York by Oswald T. Avery, Colin M. MacLeod, and Maclyn McCarty, in 1944, to determine the molecular constituent that carried the hereditary message. They proved that this constituent was deoxyribonucleic acid (DNA), and not other substances (3). Then ensued the frantic race to identify the configuration and the mechanism of synthesis of DNA, the ultimate goal of the first experiments of molecular biologists. Most molecular biologists have never been interested in bacteria as such. Their approach was comparable to that of an earlier generation of geneticists who had collected a very rich mass of data on the morphological and physiological genetic traits of *Drosophila* without showing any interest in this insect or in entomology in general. Most molecular biologists concentrated on the bacterial cell and took for granted the concept of isolated bacterial species. They ignored cooperative associations of bacteria. During this period several new antibiotics were discovered. Basic microbiologists used them as research tools while the medical group exploited their therapeutic applications. There was, however, little communication between these two categories of bacteriologists.

The inestimable progress that continues in molecular biology probably will permit, in the long run, an understanding of the physico–chemical basis of the majority of bacterial activities. However, most of the advances are not likely to be achieved for two or three scientific generations. In the meantime, other methods of studying bacteria allow us to progress equally in the field of bacteriology and its many subspecializations. They are equivalent to methods used in zoology and botany, fields that had already developed their own style before the advent of molecular biology.

Recognition of Genetic Unity in Bacteria

In recent years, there has been increasing evidence for the existence in bacteria of capacities that are more complex than those that could be attributed to rather simple species. Recognition of the genetic unity of bacteria, at first tacit, is now explicit and becoming more and more widespread. This concept of genetic unity is based on the possibility that in nature any bacterium has access to most, and possibly to all, genes belonging to other bacterial strains. This access more than compensates for the small number of genes existing in any bacterial cell—such a small number that the cell might be considered to be incomplete.

> *Common information and communication molecules available to all bacteria. No genetic isolation in bacteria, thus absence of true species. A common potential genome corresponding to a planetary bacterial entity.*

Today bacteriologists have doubts about using the traditional Linnean classification of bacteria. Initially, new classification concepts developed from the work of bacteriologists interested in taxonomy. They demonstrated that new information on the generalized exchange of genes among bacteria substantiated the proposition that bacteria are not subject to genetic isolation and thus cannot be considered to be divided into true species (16,18,23,26,42,44–46). The uniqueness of the genetic system of prokaryotes and the general consequences that follow from it were stressed (1,8,10,13,16,18,23,31,33,34,38,40,42–44,48,54). Subsequently, the widespread development in nature of transmissible hereditary resistance to antibacterial drugs proved irrefutably that bacteria act as a united entity capable of solving complex problems, and solve them very efficiently every time (17,57).

We have become advocates of the concept of a common potential genome (the entire hereditary reserve of one biological entity) in bacteria and of its logical extension, the notion of a single planetary entity comprising all bacteria (9,27,47–51). It is our contention that this entity is a unique, complex type of clone composed of highly differentiated (specialized) cells, and that this type of clone is exclusive to prokaryotes. We have also compared the complex functions of the bacterial superorganism working for the general well-being of the entire biosphere to the functions of a computer with a large data bank and an extensive communications network (47). This analogy allows us to make a coherent description of the bacterial world, reflecting its central role in life on Earth. Epidemiologists were among the first to grasp the importance of the concept of a global entity of bacteria. They were confronted by a striking demonstration of unity among different strains of bacteria that helped each other resist antibacterial drugs. A kind of social behavior, comparable in some ways to that of social insects or humans, appears to be exhibited in the bacterial world. The binding force among bacteria is one of constant exchange and permanent choice of information bits, principally genes. Recent developments in genetic engineering must not obscure the fact that in nature bacteria by themselves can and do exchange genes at all times and between different types of strains. Eukaryotes, however, carry out chromosomal exchanges by means of sexual reproductive processes and between members of the same species. In the laboratory, some viruses can be induced to transfer a few eukaryote genes, but the extent of eukaryotic gene exchange in nature, although shown to be possible, has not yet been proved (2,8,37,41). Thus, bacteria have revealed that they form a highly efficient, complex, unified entity whose constituent elements have great autonomy, and that this entity is unique in the biosphere. Because of their originality and exceptional importance, bacteria are finally, like Cinderella, being recognized as attractive and respectable organisms.

CHAPTER 2

Organization at the Cell Level

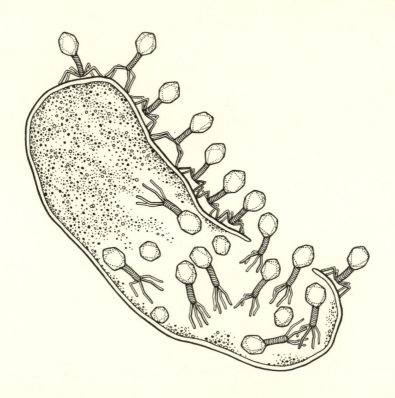

MOST OF THE discoveries about bacteria have been realized at the cell level. In Chapter 4 we will consider important associations of these social microbes. In this chapter we turn our attention to some aspects of the individual bacterial cell.

Structure

The general shape of a bacterial (prokaryotic) cell usually corresponds to one of three main types: rod-shaped (cylindroid), spheroid, or spiral (Figure 2–1). In nature, bacteria are usually found in mixed populations of rod-like and spherical forms (Figure 2–2). In addition, another type lacking a fixed form, because it has no rigid membrane, is found in mycoplasmas, a group of bacteria that permanently lack a cell wall, and in L forms, revers-

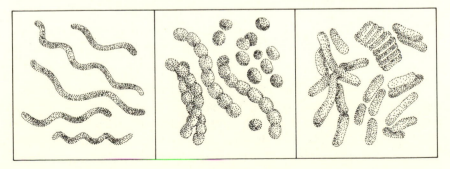

Figure 2–1. The usual forms of bacteria (left to right): rods, spheres, spirals.

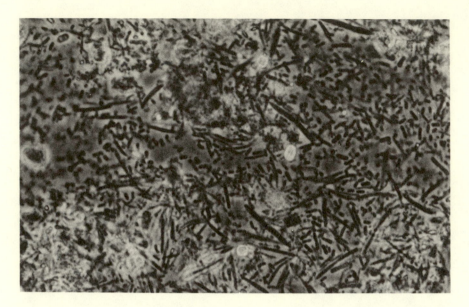

Figure 2–2. Micrograph showing mixed bacterial teams in mouse intestinal content. (×2000, fuchsine col. Courtesy of G. Nogrady, University of Montreal.)

ible variations of different bacteria, temporarily lacking their cell walls. There are a few bacteria with a more complex aspect but they are still structurally simpler than eukaryotes. This relative uniformity of morphological types in bacteria is very surprising after three billion years of evolution. There is some indication that strong competition eliminated less favorable forms. In any case, this uniformity reinforces the idea of unity in the bacterial world.

> The bacterial cell: limited to a few morphological types and a minimal size

The smallest diameter found in bacterial cells is also relatively uniform, approximating one thousandth of a millimeter, or one micrometer (μm). The center, or geometric axis, of the cell is thus very close to its external surface, permitting easy and rapid exchange of nutrients with the environment, as well as rapid elimination of wastes from vital processes.

A few bacterial strains produce certain cell forms specialized for resistance to unfavorable environmental factors: these are called *spores*. The most typical form, the endospore, is found mainly in bacterial groups known as genera named *Clostridium, Bacillus,* and *Sporosarcina*, and probably in *Coxiella*. Only one endospore is formed in a bacterial cell. Endospores are the living cells by far the most resistant to aging and to the harmful effects of physical and chemical substances. In another large group of bacteria, the actinobacteria, multiple external spores (exospores) are found at the tip of filaments that are usually aerial; these spores are less resistant to extreme environments than endospores.

Essential differences between prokaryotes and eukaryotic cells

Bacterial ultrastructure, as revealed by electron microscopy and constituent analysis, differs markedly from the ultrastructure of eukaryotic cells (Figures 2–3 and 2–4 and Table 1). The great majority of bacteria possess a rigid cell wall on the outer side of the cytoplasmic membrane. This entire wall or, in some cases, its internal surface layer, is composed of a type of macromolecule found only in prokaryotes: *peptidoglycan*, sometimes called *murein*

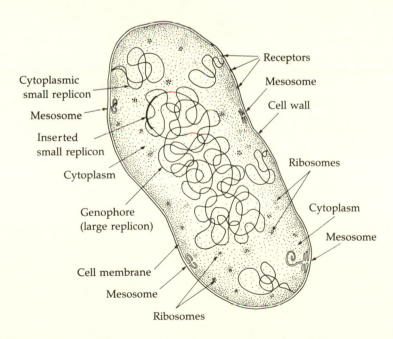

Cytoplasmic small replicon

Mesosome

Inserted small replicon

Cytoplasm

Genophore (large replicon)

Cell membrane

Mesosome

Ribosomes

Receptors

Mesosome

Cell wall

Ribosomes

Cytoplasm

Mesosome

Figure 2–3. Diagram of prokaryotic cell structure.

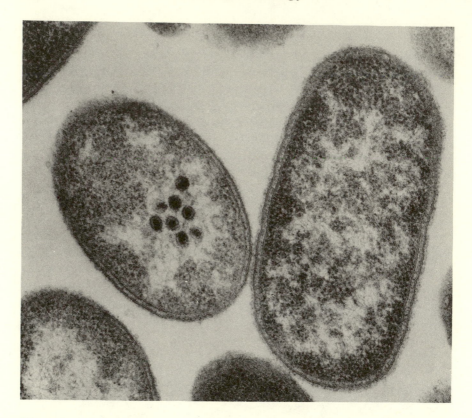

Figure 2–4. Micrograph showing bacterial cell structures. The dark spots in the cell on the left are phages in the cell. (*Bacteroides fragilis*; ×66,000, Courtesy of David Chase, Veterans Administration Hospital, Sepulveda, California.)

or *mucopeptide*. This rigid substance maintains the form of the cell. Its principal role is to support the cytoplasmic membrane, so that the cell does not explode as a result of the very high osmotic pressure of the cytoplasm. The concentration of molecules in the cytoplasm, which is much greater in bacteria than in eukaryotes, contributes to a very rapid growth in bacteria. The cell wall owes its rigidity to a chain-link molecular configuration of fibrils of polysaccharide (a polymer of N-acetyl-glucosaminic acid and muramic acid) running in one direction and short peptide chains making perpendicular connections. The chemical composition of the peptidoglycan varies very little among strains; the only difference is in the level of its constituent peptides. This homogeneity again underlines the unity of the prokaryotic world. Only the cell walls of Archaebacteria lack muramic acid (5,59). This group of bacteria, which includes the methanogens, seems to have kept many ancient properties, some of them suggesting a relationship with those ancestors of the eukaryote providing the cytoplasm and the cell membranes. Peptidoglycan is the target of many anti-bacterial drugs or antibiotics, such as penicillin, that are not toxic to eukaryotes since eukaryotic cells do not contain this cell wall constitutent.

The cell wall, a composition exclusive to bacteria. More complex in Gram-negative bacteria and favorable to genetic exchange between very different strains

The composition of the bacterial cell walls determines whether a strain will show the color violet (Gram-positive) or red (Gram-negative) after being stained with the Gram stain. The cell wall of Gram-negative bacteria contains, in addition to the peptidoglycan layer, an outer lipid layer composed of highly resistant glycolipid and lipoproteic complexes. There are spaces between these layers where enzymes called *exoenzymes*, secreted in the liquids surrounding the cell, attack food of high molecular weight that cannot cross the cytoplasmic membrane. Moreover, the complexity of their cell wall allows Gram-negative bacteria to perform conjugation, a very effective type of genetic exchange that can take place between markedly different strains. These exchanges occur by means of physical contact between cells with the help of conjugation pili, long proteinaceous tubules projecting outwards from the cells of Gram-negative bacteria possessing conjugation plasmids. (Genetic exchanges will be discussed in detail later in this chapter.)

Bacteria that can move in liquids and dilute gels also bear on their surface long mobile filaments (flagella) composed of proteins called flagellins. The number and location of flagella vary among different groups of bacteria.

In some bacteria, the cell wall is surrounded by the *capsule*, a mucilaginous layer with a smooth outer surface most frequently made of polysaccharides, although in rare instances it may be composed of peptides. In other bacteria, the mucilaginous layer has a rough exterior that can adhere to an irregular surface. This type of layer, called *slime*, plays an important part in adherence phenomena.

In the morphologically more complex bacteria, especially Actinobacteria, cells can remain attached to one another and form structures with primitive branches. They can differentiate as microscopic threads (hyphae) growing upwards from the surface on which they multiply and bearing multiple aerial spores (exospores). As already mentioned, exospores are less resistant than, and formed differently from, endospores, where only one spore is produced per cell.

> *Numerous surface receptors on bacteria for various substances, particularly information and communication molecules*

Most bacteria possess surface structures that act as receptors for various substances, including information and communication molecules such as DNA (Figure 2–3). Under the cell wall and enveloping the cytoplasm there is a cytoplasmic membrane composed of a mosaic of proteins and lipids, as in eukaryotes. However, bacteria differ from eukaryotes in that, with a few rare exceptions such as in mycoplasmas, the cytoplasmic membrane in bacteria contains no sterols. These substances are not synthesized by bacteria. This property causes the cytoplasmic membrane in bacteria to be sensitive to certain antibiotics that are less toxic to eukaryotes. The cytoplasmic membrane of bacteria is permeable to dissolved substances of low molecular weight only. It contains the cell cytochromes, important molecules for energy metabolism, located principally in the more highly-developed folds of the membrane, called mesosomes. Each self-replicating bacterial DNA strand seems to be attached to a mesosome.

All the other substances necessary to the life of the cell are contained in the bacterial cytoplasm. The cytoplasm is primarily composed of molecules involved in nucleic acid and protein synthesis (nucleotides, messenger and transfer RNA, amino acids), substances required for energy metabolism (sugars and adenosine triphosphate (ATP)), enzymes necessary for various metabolic reactions, and salts. Ribosomes, small structures where proteins are formed according to RNA instructions, visible by means of electron microscopy, also float freely in the bacterial cytoplasm and are not bound to membranes as is the case with eukaryotic ribosomes. Bacterial ribosomes are slightly smaller than those of eukaryotes; their structure is also slightly different. Furthermore, their function differs sufficiently from that of eukaryotic ribosomes that bacterial ribosomes can be used as targets for several antibiotics specifically blocking bacterial protein synthesis.

In the center of a bacterium lies a large, double-stranded, circular strand of DNA approximately one millimeter long when uncoiled (Figure 2–3). It is not enveloped by histones as is eukaryotic DNA. Eukaryotic DNA is wrapped by five histone proteins that make a discrete knobbed thready material called chromatin. Bacterial genetic material is never chromatin, nor is it confined by a nuclear membrane. It floats freely in the cytoplasm. It is visible by electron microscopy as a nucleoid, and it contains most of the hereditary material of the cell. It is called a chromosome by many bacteriologists, but it should probably be called a genophore. Each bacterial cell contains 200 to 10,000 times less genetic material than do eukaryotic cells (4,6). In most bacterial cells there are also a few much smaller, similar DNA molecules invisible in ordinary EM microphotography; they are the self-replicating prophages and plasmids, both called small replicons (Figure 2–3). They can be transmitted to other strains. The prophage copies are protein-coated virus-like particles (the phages, also called bacteriophages) able to persist a few months in nature. Plasmids spread actively only by physical contact between two bacteria.

Genetics

Generalities. The most striking differences between prokaryotes and eukaryotes are in their genetic organization. Bacteria are able to multiply and live in liquid or gelified exposed environments, competing with other organisms or associating with them and reacting very rapidly to selective pressures. Because of the extreme rapidity of bacterial multiplication, a single bacterium experiencing relatively favorable conditions can produce a multitude of descendants in a very short time, using relatively small quantities of nutrients. The short generation time and the very high numbers of bacteria under the right conditions enhance the probability of the appearance of usually rare genetic modifications.

Because the principal intracellular bacterial functions are controlled by enzymes, which are proteins under direct genetic control, the genetic system is usually linked directly to all bacterial functions without the intervention of other regulatory mechanisms. Most of the genes in a bacterial cell are structural genes, as they each code for the structure of a particular protein. They may be inactivated by an operator gene that helps regulate the activity of the adjacent structural gene (or a series of genes) that should not be active when substances called repressors are accumulating. A repressor binds with the operator gene and this bound gene inhibits the neighboring structural gene (or a series of genes), called an operon if it is under the control of only one operator gene. The repressor is synthesized under the guidance of a regulator gene. An operon is transcribed as a single unit, forming a RNA molecule that contains information for more than a single gene. A mutation in one of the genes close to the operator will usually stop the translation of all genes in the operon that are translated immediately after. We call these mutations polar. They show that the ribosome is unable to reinitiate translation of an operon; it cannot continue to make the proteins under the direction of the messenger RNA.

The mechanism of scissiparity (doubling into two identical cells) enables every bacterial cell that possesses genetic resources valid for its given environmental conditions to multiply without the assistance of another cell of the same species. This binary fission (asexual reproduction) produces two identical cells, perpetuating the original model and ensuring its stability, and when a hereditary modification happens in one cell it is equally transmitted. Because bacteria are "haploid"—that is, have only one gene for each property—bacterial cells usually carry no recessive genetic traits. This characteristic generally permits the rapid phenotypic manifestation of any newly-acquired hereditary property. Most of the important modifications in these hereditary properties are obtained in bacteria by exchange of genes.

Separation of hereditary material. In bacteria, as in eukaryotes, heredity has two basic functions. The first is the transmission of identical information pertaining to an organization of living matter that has already proven itself in an organism or species. This function secures stability and a faithful copy in the offspring cells. The second function of heredity is to provide the possibility of improving or modifying the transmitted information in order to exploit new favorable conditions or to survive new adverse conditions. This function facilitates variety and new solutions (evolution or temporary adaptation).

In a bacterial cell, these two genetic functions are carried out by genes grouped in two different types of self-replicating DNA units, called replicons. The distribution of genes in one type or the other is adapted to existing conditions (43). If a gene proves itself necessary over a long period of time, it may change company and jump from a temporary visiting group to a more sedentary assemblage of genes. The reverse might happen too. In the few bacteria well studied genetically, all essential hereditary information required for survival and replication is contained in a large replicon identical to, or constituting the major part of, the large DNA molecule or genophore (incorrectly called the chromosome) that forms a filamentous closed loop located at the center of the cell. It is much larger (in most cases, 2.5×10^9 daltons) than the other circular DNA molecules present in most bacterial cells. Its exclusively essential hereditary information is maintained at a minimum, with an exceptional limitation of functions, enough for the cell to reproduce itself and to perform its usually very specialized metabolism (47). With very few exceptions bacterial genes are not repressed during each generation, although very few are temporarily repressed when necessary. Because of its reduced size, the bacterial intracellular genome, and in particular its genophore, is incomplete for many important circumstances. Thus, a reserve of repressed genes is not available in the cell for possible tissue and organ differentiation, as in eukaryotes. The large replicon does not possess information for its own active transfer to another different cell.

Had bacteria possessed only the type of hereditary material present in the genophore, as was and still is partially believed, their world would have been a particularly static and inadaptable one. There would not have been much room for evolutionary innovation or for differentiation. Extinction very probably would have been a serious threat to many specialized bacteria. Fortunately for bacteria, temporary accessory genes visiting from one sharing strain to another are also present in most bacteria (see Table 2). Most are grouped in self-replicating DNA units called

Table 2. Types of Genes Exchanged by Bacteria

DNA fragments (incapable of self-replication and active transfer)

Must be integrated into the length of a replicon in order to divide; often carried passively to a different cell; genes usually derepressed.

	Usual DNA fragments	*Transposons*
	Devoid of any special capacity to integrate into a new replicon; must find a zone of homology and recombine in that region of a replicon in a recipient cell.	Have insertion sequences at their ends that enable them to be inserted into the length of any replicon in their own or a recipient cell.
Methods of transfer		
Active (coded by their own genes)	—	—
Passive (by soluble DNA, only in immediate vicinity)	Transformation	Transformation
Passive (by small replicons)	Conjugation Transduction	Conjugation Transduction
Metabolic range of transfer	Related types only	Different types
General structure		
Active transfer genes		
Conversion genes		
Insertion sequences		
Replication genes		

Small replicons (self-replicating DNA molecules other than the genophore)

Circular molecules, at least for their autonomous replication; consist of replication genes plus derepressed conversion genes that might help the recipient cell; may be lost when there are no favorable selective pressures for keeping them; attach to the cell membrane or insert into the length of a replicon, usually the genophore (a few, called episomes, may be able to do both); have some similarities to DNA virulent phages and DNA eukaryotic viruses.

Nonself-transferable small replicons	Self-transferable small replicons	
Plasmids or defective prophages; usually attach on the cell membrane.	Their genes for self-transfer (usually repressed) also code for organelles such as pili or heads and tails of phages, and for important rearrangements of macromolecules in donor cells. In active transfer they need specific receptors at the surface of recipient cells.	
	Conjugation plasmids	*Prophages*
	Attach on cell membrane more often than in length of genophore	Attach in length of genophore more often than on cell membrane
—	Conjugation (needs physical contact between cells)	Lysogenization and transduction (distance between cells varies from near to far)
Transfection	Transfection	Transfection
Conjugation Transduction	Transduction	Conjugation
Most bacteria	Practically all Gram-negative bacteria	Related types only

small replicons (43). The small replicons include plasmids, episomes, prophages, phages (bacteriophages), and viruses. All are formed by double-stranded, circular DNA, as is the genophore. Some small replicons are physically separated from the large one, and other types are inserted in the length of the genophore, usually dividing synchronously with it but also able to start self-replicating. Episomes are small replicons able to appear in one or another of these forms, either physically separated from or integrated into the large replicon. The larger of the small replicons divide generally at the same rate as the cell and its genophore. The smaller ones have a tendency to make more than one copy per cell, and the smaller the DNA molecule, the higher the number of copies. In addition to the genes that enable them to multiply, most of the small replicons contain conversion genes. Conversion is the mechanism by which the host cell expresses physically information brought to it by visiting genes from small replicons. Conversion is done mostly by synthesizing new enzymes. When the conversion is without advantage for the host strain, the small replicon responsible for it will slowly disappear by dilution. This disappearance of small replicons is called curing, because when the phenomenon was discovered the small replicons were considered to be infectious elements. Today, we know that their role is to offer variation to bacteria. Each small replicon may visit thousands of different bacterial strains and each bacterial cell, although usually harboring only a few different small replicons at a time, is able to be visited by tens or hundreds of different types.

Many small replicons are able to realize the active transfer of one of their copies to other bacterial strains. These self-transferable small replicons have many common characteristics. The genes coding for their transfer are usually repressed by a protein molecule, the repressor that also acts on their autonomous replication. The active transfer of small replicons needs the synthesis of some specific proteins, structural and enzymatic. Very often the self-transferable small replicons not only transmit their own genes but they may equally transfer genes from another replicon (either large or small) or even an entire nonself-transferable small replicon. In self-transferable small replicons, the genes not responsible for their reproduction or transfer are derepressed and are active in the conversion of the visited cell. In this way, the visited cell usually acquires a new capacity to synthesize different proteins, mostly enzymes or surface receptors.

The most wide-spread type of self-transferable small replicon is the prophage (8,21,31,42). (See Table 2.) Prophages are present in Gram-positive and Gram-negative bacteria, in cyanobacteria (50), and other bacteria. Most of them are inserted in the length of the genophore, where they maintain a delicate autonomy. Usually they divide simultaneously with it but in many circumstances they start multiplying independently. A few are permanently separated from the genophore and float freely in the cytoplasm; even fewer are able to be real episomes and thus switch from being part of the genophore to being free floating. Under usual circumstances a very small minority of the prophage-containing cells of a lysogenic bacterial strain will have their prophage overcome its repressor and realize the steps for self-transfer. This phenomenon is called *induction*. As a result of induction, tens or hundreds of prophage copies will be synthesized in the induced cell, and they will be covered with a protein coating and then liberated outside the destroyed cell so that it is possible for each protein-coated prophage genome (a temperate phage) to be transmitted to a bacterium from another strain. The proportion of induced cells, and therefore of synthesized phages, increases under many unfavorable circumstances; experimentally it can be produced by ultraviolet radiation, many chemical substances, and other means. The liberated phages may persist in nature and travel widely by means of rivers, sea currents, or wind. Prophages can be exchanged only between related strains of bacteria that have similar metabolisms. However, by means of intermediate strains or phages, in the end a very extensive gene exchange may be realized by prophages (Figure 2–5).

The other self-transferable small replicons are the conjugation plasmids (see Table 2). A few of them are inserted in the genophore but most of them are autonomous, and very few are episomes. If their genes for transfer are derepressed, they may be transmitted and then the protein involved in the transfer is synthesized to produce one or two pili (long ultramicroscopic tubes, protruding at the surface of the cell, that prepare to transfer a copy of the plasmid). For conjugation, the pilus has to be attached to the recipient cell, so conjugation plasmids need physical contact between the two cells realizing their transfer. Under very

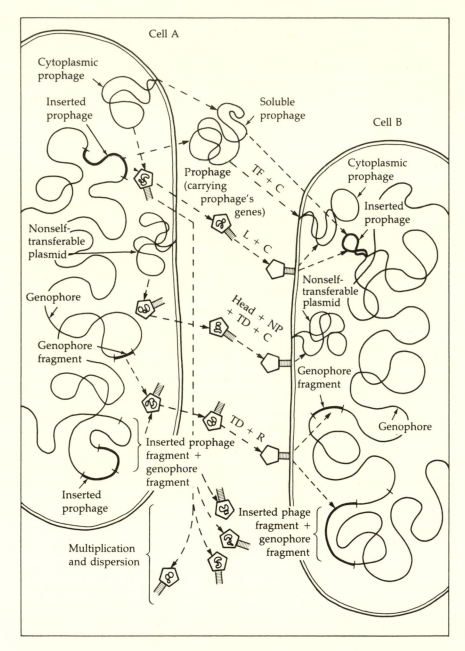

Figure 2–5. Ways in which a prophage may transmit genes from one bacterium to another. NP = nonself-transferable plasmid, C = conversion, L = lysogenization, R = recombination, TF = transfection, TD = transduction.

particular experimental conditions, some conjugation plasmids may transfer the entire genophore of the donor cell. As this transfer is always realized in the same length of time and starts from the same genes, this method is the basic one for mapping the succession of genes in Gram-negative bacteria, the only bacteria possessing well-studied conjugation plasmids. In contrast with prophages, conjugation plasmids can be exchanged between Gram-negative bacteria of very different metabolic types.

Nonself-transferable plasmids exist in most types of bacteria, both Gram-positive and Gram-negative. Their genes for self-replication or for conversion are permanently derepressed. They are transferred passively by different methods, including the help of prophages or of conjugation plasmids.

Most of the small replicons that are not inserted in the length of the genophore are attached at one point to the cytoplasmic membrane. The membrane has a limited, specific number of attachment sites for different types of replicons.

Another type of accessory visiting gene, different from the small replicons, are transposons, small DNA segments comprising a few genes flanked by insertion sequences that allow them to jump from one replicon to another in the same bacterial cell. Transposons can probably insert themselves in any replicon of a new cell in which they are passively introduced (see Table 2).

Small replicons and transposons are groups of bacterial genes shared by many diverse bacterial strains. They may even pick up genes from other replicons from the cells they visit and transfer them to different strains, thus allowing all possibly useful genes to be offered to all possible bacterial hosts.

Hereditary changes

Mutations. The normal mutation rate in bacteria is not significantly different from that in other living cells, being of the order of 10^{-6}–10^{-9} per cell per generation. Most bacterial mutations occur through the loss of gene function, representing the loss of the capacity to synthesize an enzyme. This type of mutation generally constitutes a drawback for the bacterium and leads to its death, although in rare cases loss of or change in gene function may be beneficial. As bacteria reach very high numbers in nature, favorable mutations for given environments do appear.

Resistance to different chemical or biological agents is often the result of some very simple mutation in bacteria. However, the arrival of a new, favorable gene from another strain is more often the decisive solution in many challenging situations. Bacterial mutants are extensively used in research. There is a proportionally equivalent frequency of mutations on a large or on a small replicon, depending on their size.

Genetic recombination. Recombination is a rearrangement of sequences of genetic material such that a new linear sequence is produced from more than a single source (parent). It may occur by exchange of similar (homologous) fragments, or by the integration of insertion sequences, transposons, or small replicons (prophages or plasmids). Many recombinations produce mutations, permanent hereditary changes in the DNA sequences. Some transfers of prophages or of plasmids need recombination in order to be integrated into one of the recipient cell's self-replicating DNA molecules, usually into the genophore.

Mechanisms of gene transfer. There is in nature a surprising variety of gene transfer mechanisms from one bacterial strain to another, permitting the rapid acquisition of genes that may become beneficial in changing environments. Individual cells of many strains may benefit from more than one of these mechanisms (Figure 2–6).

Transformation is a phenomenon by which genetic material is transmitted from one bacterial strain to another in the form of soluble fragments of DNA, originating from a live or dead cell. The DNA fragments are dissolved in the external medium and subsequently penetrate a bacterial cell of another strain. In general, DNA can only penetrate cells that have receptor sites for DNA on their surfaces. Such cells are termed "competent," and a single competent bacterium possesses approximately 10 to 20 penetration sites with favorable receptors. A substance called an activator can confer competence to, or increase the competence of, either the cell that has synthesized it or of the neighboring cells. The DNA fragment that has penetrated a cell has to be inserted into a replicon, usually the genophore, in order to be accepted and copied. The fragment it usually replaces, by recombination, is a short section of the receptor cell's DNA that contains a zone of homology (a similar section).

This mechanism, in which only a small number of genes is

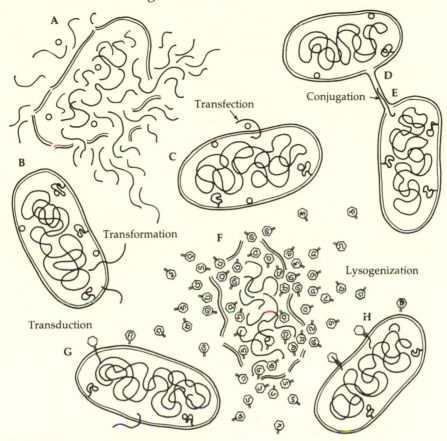

Figure 2–6. Diagram of bacterial gene exchanges in nature.

A: Dead cell liberating DNA fragments, including a few circular small replicons.

B: Cell receiving DNA fragments from cell A. One fragment finds a zone of homology on the genophore and will integrate into it (transformation).

C: Cell receiving a small replicon from cell A (transfection).

D: Cell possessing conjugation plasmids and transferring them to cell E with the help of a tubular pilus (conjugation).

F: Cell in which many temperate phages were produced. The cell lysed and the phages were dispersed.

G: Cell injected by a phage from cell F containing a fragment from the genophore of cell F. This fragment has found a zone of homology on the genophore of cell G and will integrate into it (transduction).

H: Cell injected by a phage from cell F, with its prophage genome, which is integrating into the genophore of cell H (lysogenization with conversion).

transferred, occurs in nature with a low frequency, generally of the order of 10^{-6}, although in exceptional experimental conditions the frequency may approach 5% of cells exposed to a concentrated solution of DNA. Genetic exchanges by transformation generally take place between cells belonging to strains of the same or similar metabolic types. The main barrier to transformation is the cell wall that presents openings for DNA only in competent bacteria. In bacteria with altered or incomplete cell walls, or in those that lack cell walls altogether (protoplasts, derived from Gram-positive bacteria; spheroplasts, derived from Gram-negative bacteria; L forms; mycoplasmas), it seems to be easier to induce transformation. Another barrier to transformation is the restriction enzymes that partially block the acceptance of foreign DNA; their action is partially countered by a modification enzyme that usually is also present.

Transfection is a mechanism similar to transformation, in which the soluble DNA molecule that penetrates and is accepted by the bacterium is an intact small replicon (a prophage or a plasmid) originating from another bacterial cell, the donor. It attaches itself to the same type of site in the receptor cell as the normal infectious form would occupy. Transfection does not necessarily involve recombination and consists simply of the addition of a small shared replicon to the recipient cell (Figure 2–6).

Lysogenization is a mechanism that involves the transfer of a prophage from one bacterial cell to another. In nature, this transfer occurs very rarely by transfection. In nearly all prokaryotes, it occurs commonly through the intermediary of an infectious form of prophage, the temperate bacteriophage, or phage. In a lysogenic strain—that is, a strain harboring a prophage—the majority of bacterial cells multiply normally and the prophage replicates in synchrony with the bacterial genophore in which it is sometimes inserted. Such a strain will contain a very low proportion (10^{-2} to 10^{-4}) of cells that allow the prophage genome to undergo rapid multiple replication in preparation for its potential transfer. The prophage transfer genes control the synthesis of polymerases for the replication of the prophage genome, the first in the transfer process. This synthesis is followed by the synthesis of constituent proteins: "heads" (box-like assemblies destined to protect the genes of a prophage during transfer) and "tails" (tubular forms attached to the heads) of the future phages; then the synthesis of assembly enzymes governing the formation of complete phages occurs. For active transfer, the prophage genome must be enclosed in the protein head. This complex multiplication of the phages is followed by bacterial cell dissolution and the release of countless infectious phages. The whole phenomenon is called *natural* or *spontaneous induction*. Temperate phages released in this way can inject their genome into bacterial cells of any strain possessing receptors to which the phage tails can attach. Once inside the sensitive cell, the prophage genome can either find an attachment site and remain there as a new prophage (lysogenization), or it can direct the cell to produce infectious phages identical to the infecting temperate phage (31). A bacterial cell that has just acquired a prophage is said to be lysogenized (Figure 2–6).

The main new characteristics of a lysogenized culture are determined by the conversion genes expressed from the prophage that it harbors. This set of modifications is known as *lysogenic* or *phage conversion*. The most common phenomena are as follows. All lysogenic strains exhibit a property called *immunity*. It bears no direct relation to the phenomenon of immunity in vertebrates, but rather it is the specific resistance of lysogenic cells to the multiplication of phages with a genome identical or similar to that of the already present, visiting prophage. Immunity is caused by the repression of those prophage genes that govern its active transfer. This is effected by the repressor, a protein that is the first substance to be synthesized after the arrival of a prophage in a cell, in accordance with the information contained in a prophage gene. The repressor prevents the synthesis of polymerases required for the multiple synthesis of the prophage genome and also that of protein constituents involved in prophage transfer. As a result, the prophage divides synchronously with the bacterial genophore. This immunity also differs from resistance to phages. Resistance to phages is due to the absence of specific receptors for certain phages on the surface of bacteria.

Other consequences of lysogenic conversion are: the synthesis of several bacterial toxins composed entirely of protein, in many pathogenic strains (21); the synthesis of enzymes that vary according to the type of prophage; and the synthesis of certain molecules that form phage receptors on cell wall surfaces. Curiously enough, phage conversion almost never confers resistance to antibiotics.

The mechanism of phage conversion has been extrapolated to form the basis of one of the principal theoretical models explaining the viral origin of cancer. Some eukaryotic cells in animal tissue, which demonstrate a phenomenon analogous to bacterial conversion provided by a virus, are thought also to harbor new small pieces of replicating DNA. These cells, which grow rapidly, piling on top of each other, are said to be "transformed."

Transduction is another important prophage-controlled mechanism of genetic exchange between different bacterial strains. It is based on the ability of temperate phages to assemble in a manner such that any fragment of the bacterial cell's DNA can be inserted in the protein head in place of part or all of the prophage genome. This DNA fragment originates from the cell containing the prophage and cannot be much larger than this prophage. The modified phages formed in this way can inject DNA from any of the cell replicons into a sensitive bacterium. Once injected, DNA fragments that are not small replicons must find and replace by recombination a homologous zone on a bacterial replicon, usually the genophore. All lysogenic strains release a large number of phages by spontaneous induction, many of which carry fragments of DNA from other replicons in place of the usual prophage genome (Figure 2–5). If these genes, once they are successfully transmitted, constitute a useful replacement for the genes previously found in the homologous zone, the recipient cell benefits from transduction (Figure 2–6).

It is easy to imagine many circumstances in which transduction plays a spectacular role. For example, consider the case of a natural marsh environment, where one of many lysogenic strains has very nearly exhausted its source of nutrients. However, if it possessed the appropriate enzyme, it could utilize an abundant new source. The temporarily unfavorable conditions give rise to an increase in the incidence of spontaneous induction, and a very large number of phages is released into the environment. Hundreds, or perhaps thousands, of phages among the millions thus liberated may reach another marsh where they infect several cells of a bacterial strain that is physiologically similar to the original lysogenic strain and is sensitive to these phages. In most of the infected cells, the phage DNA would not insert itself as a prophage because there would be insufficient time for the repressor to be synthesized before a lytic cycle was triggered by the expression of the prophage genome transfer genes. Some of the numerous phages released in turn by these sensitive cells would immediately inject other cells of the same strain, or of slightly different ones, soon producing probably several million phages to be released. A certain portion of these phages would then contain genes from the replicons of a sensitive strain, which might include a gene that could be of great help to the lysogenic strain that originally released the phage. If a sufficient number of phages carrying this gene returned, carried by water or insects for example, and injected it into one or more cells of the original strain, transduction would take place with beneficial consequences for the strain that previously faced difficulties, as this cell will multiply and simultaneously transmit the useful gene to other cells.

In addition to transmitting fragments of bacterial replicons by transduction, prophages can, by the same mechanism, transmit intact small replicons such as other prophages, conjugation plasmids, and—most often—nonself-transferable plasmids. The frequency of this type of transduction is greater than 10^{-6}, the figure for the frequency of transduction involving replicon fragments, since recombination is not required when an entire small replicon is transferred. Once injected, the small replicons must simply find an attachment site, either on another replicon or on the cytoplasmic membrane. In this type of transduction, all the genes of a shared small replicon are acquired by the receptor cell (Figure 2–5). It is this mechanism that is responsible for the transmission, by nonself-transferable plasmids, of resistance to various antibacterial drugs in Gram-positive bacteria, particularly Staphylococci. Thus it can be seen that nonself-transferable plasmids can be transferred to other strains by transfection and transduction, and participate significantly in the system of genetic exchange effected by bacteriophages.

Transduction is seen in all types of bacteria, Gram-positive and Gram-negative, and clearly demonstrates the important biological role of prophages in carrying out the exchange of genetic material between various bacteria. Although the frequency of transduction is only 10^{-6}, it is significant because of the very large numbers of bacteria found in most environments.

Transduction is used in research for the detailed genetic mapping of bacteria.

Conjugation is one of the most important mechanisms of genetic transfer between different bacterial strains. It differs from the mechanisms of transformation, transfection, lysogenization with conversion, and transduction in that it cannot take place at a distance. Cells undergoing conjugation must make direct physical contact, helped by the microscopic tube called the pilus that serves in this transfer of genes (Figure 2–7). The overwhelming majority of observed cases of conjugation have involved Gram-negative bacteria. Unlike other mechanisms, conjugation can take place between strains having very different physiological characteristics.

The self-transmissible small replicons responsible for bacterial conjugation are generally known as conjugation plasmids. The most extensively studied of these plasmids have been the R plasmids, because of their role in the transmission of a so-called contagious multiple-drug resistance among pathogenic bacteria (57). In conjugation, a copy of the plasmid goes from the donor cell into the recipient one (Figure 2–6).

Similarities between plasmids, prophages, and viruses. There are many similarities between conjugation plasmids and prophages. Both are self-transferable small replicons. Both normally divide synchronously with the bacterial genophore, that is, at the time of cell division. The genes that govern the initiation and succession of the stages of their particular transfer mechanisms are normally repressed. Efficient mechanisms exist for the reversal of this repression when the cells receiving or already containing these small replicons benefit from them and transfer is favorable. Through their mechanisms of transfer from one strain to another, both conjugation plasmids and prophages can carry along a portion of the large replicon or of a small one, or an entire small replicon, particularly nonself-transferable plasmids. In the case of conjugation, the portions of the replicon are carried along by facilities of recombination between the two plasmids.

Frequently, a nonself-transferable plasmid may be temporarily inserted lengthwise in a conjugation plasmid, and this effects its relatively common transfer in the process called "mobilization." As we have seen, nonself-transferable plasmids are also transferred by transduction, the mechanism of genetic transfer that is effected by prophages. Thus, the three types of small repli-

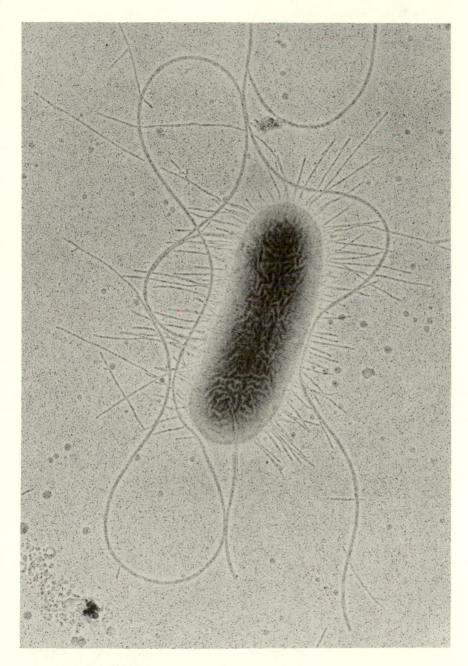

Figure 2–7. Micrograph of *Escherichia coli* showing conjugation pili (thin, almost straight projections). The long curved structures are flagellae. (×25,500. Courtesy of David Chase, Veterans Administration Hospital, Sepulveda, California.)

cons complement each other in carrying out the exchange of genes among bacterial strains.

Insertion sequences and transposons (sometimes called "jumping genes") favor the ideal redistribution of genes in a bacterial cell at all times. Genes that are most useful on a long-term basis can thus shift from highly transitory small replicons to the much more stable genophore. Conversely, genes that are rarely used can shift from the genophore to a small replicon. Thus, the small replicons and the genophore can form favorable assortments of intracellular genes from the great reserve of genes accessible to all bacteria. In addition, different genes can be integrated for varying periods of time in special associations according to the temporary needs of any given strain (12). It is very probable that transposons may be transferred from one strain to another by conjugation, just as they may be transferred by transduction and by transformation. Transposons do not need zones of homology like the usual DNA fragments and could be inserted in any replicon of the recipient cell.

All these genetic facts are known and accepted in bacteriology. Bacterial transposons are a discovery of the last decade (12). However, all this knowledge has almost always been considered piecemeal, even in recent textbooks, and thus its significance has been grossly underestimated. As the temperate phages were discovered long before their small replicon origin, the prophages, the latter were never accepted by most bacteriologists for what they are: the most important type of visiting genes, able to relate many different bacterial strains to each other even when they are separated geographically. Prophages are still looked at uneasily as some kind of irregular form of a virus. All the phages are still wrongly seen by many biologists as bacterial viruses, although the temperate ones are virus-like transfer forms for a very widespread normal hereditary component of bacteria. Viruses in general are similarly formed by one protein-coated nucleic acid molecule (DNA or RNA), and they have genes instructing the sensitive cells how to synthesize identical virus particles. Viruses are usually just parasites. The few scientific papers (2,8,37,41) that try to show exceptional cases in which viruses carry favorable genes and help even eukaryote evolution have not been supported by facts. By contrast, nobody contests the positive contribution of prophages to many bacterial strains. There are equally many similarities between plasmids and viruses; however, no similar confusion happened, as plasmids were discovered nearly thirty years later than phages, and as they do not possess a virus-like transmission form.

What has been missed by most bacteriologists is the importance of the parallel activities of the shared genetic material and of complete bacterial cells. There is a continual exchange and flow of information, among small and large replicons and among bacterial cells. Everywhere in the bacterial world there is active competition and evolution among the thousands of types of prophages, plasmids, and transposons to spread into new bacteria, to replace former occupants (47,49), and to achieve the best local temporary solution. This competitive and evolutionary activity is exhibited by bacteria as well. All these elements work together to make up the bacterial world.

In soil and in the digestive systems of all animals, a bacterial cell contains, on the average, three to four different small replicons. They exert a conversion activity and many are actively engaged in the transfer of their copies to other strains. Dead bacterial cells offer their genes to neighbor strains by transformation and transfection. Because the essential genes from the genophore are incomplete in many circumstances, the constant offer of new genes helps most strains to survive at one moment or another of their life. It is a clear sign of cooperation, of genetic solidarity. The facility of conjugation between practically all the Gram-negative bacteria is generally accepted and suggests a common gene pool for this important part of the bacterial world. Moreover, many experiments suggest possible gene transfer between Gram-positive and Gram-negative strains (15,20). There are indications that in nature large groups of Gram-positive bacteria exchange genes among themselves by transformation, transfection, lysogenization, and transduction, including the participation of transposons. There might be two to twenty different gene pools for all bacteria. Indeed, it is most probable that any bacterium can donate genes to and receive genes from any other, no matter how circuitous the route. This implies only one common gene pool for all the bacteria of our planet.

The probability of direct gene exchange increases as metabolic similarities between strains increase, and is very remote for entirely different bacteria, but there are always possible intermediate receptor and broadcasting strains. In a cell, the genes localized in prophages, plasmids, and transposons are seminomadic. They may spread easily for additional strains and they may disappear in one strain and be replaced by a more useful gene grouping. The genes in the genophore are rather sedentary; however, they may be spread—although seldom are in nature—by transformation, transduction, conjugation, and also by transposons and insertion sequences.

For any individual bacterial cell, the possibility of benefiting from a favorable gene as a result of these exchanges is extremely small; however, sometimes it happens. At the level of large bacterial teams, or further, at the level of the global entity comprising all bacteria, the great investment of structures and energy for the exchange of genes in all bacteria finds complete justification (see Chapter 4). In combination with the exceptional bacterial capacity to undergo short periods of rapid selection, the exchange of genes forms the strong binding force which unifies the bacterial world, endowing it with incomparable solidarity (1,8,10,12,13,17,18,23, 26,28,29,31,38,42–44,47,50,51,57).

Spread of information at the molecular level. In bacteria, as in eukaryotes, DNA is the substance responsible for both gene replication and gene function. It consists of a double helix containing two complementary polynucleotide strands, in each of which purine and pyrimidine bases are supported by alternating deoxyribose and phosphate groups. The two strands are held together by hydrogen bonds that exist only between adenine and thymine (A–T pair) and between guanine and cytosine (G–C pair) as shown in Figure 2–8. The genophore of *E. coli* contains, for example, approximately 5×10^6 base pairs. The small replicons usually have 5×10^4 or fewer base pairs.

For DNA duplication, each strand, after separation, acts as a template on which is assembled a complementary strand. By the action of polymerases, two new DNA molecules result, both identical to the original. One of its strands is in one of the resulting molecules and the other strand is in the second one. The large replicon or a small one duplicates the same way, sequentially, down its molecule. The original replicon DNA initiates its separation and replication at a site called the replicator. The frequency of the duplication of each bacterial replicon is regulated with the

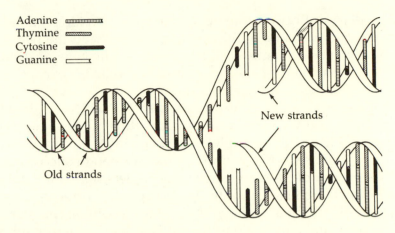

Adenine
Thymine
Cytosine
Guanine

New strands

Old strands

Figure 2–8. Diagram of DNA replication.

help of a regulatory substance, the initiator. In the bacterial cell the two resulting DNA copies and their attachment sites are separated by a localized membrane synthesis, later completed by the synthesis of the new, dividing cell wall. At the end of the duplication of all replicons in a cell, and the division of the cell in two, the exact information carried by all the replicons is spread to the two resulting cells and later to their descendants by the same process. Small replicons may divide at a higher rate than that of large replicons, and may be transmitted to other and different bacterial cells, independently of any division of the cells involved. Thus, in parallel with the spread of exactly the same information from one cell to its descendants in bacteria, each small replicon may independently spread to different cells. A permanent redistribution of genes takes place among bacteria by one of the methods we already mentioned; most of the resulting changes are short lived, however. When new genes are needed they are available, and if they spread the appropriate information they are kept by the visited cells and their copies will spread to new ones. Most of the intracellular bacterial genes code for information directly related to their cell growth, which is normally followed by cell division; most genes in any replicon code for the amino acid sequence of a resulting protein, usually an enzyme. As in all living things, protein synthesis in bacteria is the result of a typical sequence of events. The two strands of DNA separate temporarily and messenger RNA (mRNA) is formed so that its nucleotide sequence is complementary to one of these strands. Thus, with the help of a RNA polymerase the mRNA is a template of any gene or sequential genes. This part of protein synthesis is called *transcription*.

Simultanously, intracellular amino acids are bonded to one end of a specific RNA molecule, transfer RNA (tRNA), provided at the other end with the corresponding three nucleotides, complementary to one of the triplet of bases on mRNA. With the help of a ribosome, each tRNA finds in the nucleotide succession of mRNA its complementary nucleotide triplets; its amino acid is fixed by peptide linkage to the preceding one, corresponding to the preceding triplets of the mRNA. Thus, a peptide chain grows as the ribosome moves along the mRNA, and the entire mRNA molecule is translated into a coded sequence of amino acids. This part of protein synthesis is called *translation*.

Differentiation. In bacteria—with the exception of sporeforming strains, the multicellular cyanobacteria, and myxobacteria showing morphology variation—differentiation concerns exclusively a change in enzyme synthesis. Morphological differentiation is limited. Most of this bacterial differentiation is not obtained, as in eukaryotes, by the play of repression and derepression of intracellular genes that are repressed at 90–95 per cent. Instead, bacterial differentiation usually happens when new genes from another strain arrive in one cell or when a cell loses some of its genes. Most of these changes are the result of the exchanges of small replicons between different strains, and the small replicon most often present in the different bacterial groups is the prophage. Lysogenization and curing of a lysogenic cell seem to be a very widespread form of bacterial differentiation.

Each bacterium contains only an infinitesimal portion of the accessible genes from the common potential genome of all bacteria. When the exchange of genes leads to a modification of the genophore, a long-term differentiation is produced, whereas the acquisition or the loss of a prophage or of a plasmid represents a short-term differentiation. The totality of bacterial genetic material may be in a state of dynamic flux. Differentiation and redifferentiation may arise and become less temporary when they help bacteria in new circumstances. When harmful or neutral, as they are produced daily in billions of cells, these differentiated cells are simply diluted out in the mass of better adapted bacterial cells. The severe pressures of a hostile environment, as nutrients decrease and water becomes scarce, will allow only the most appropriate combinations for continued reproduction. In most cases, these will be the ones that survive, sometimes improved by a few new genes.

CHAPTER 3

Bacterial Metabolism and Its Control

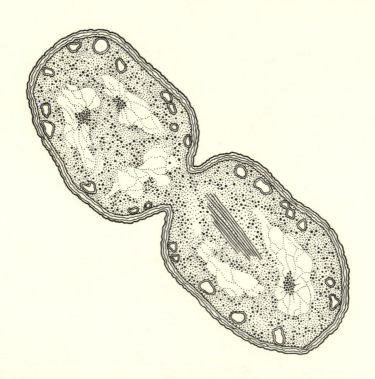

MOST BACTERIA are unicellular and require a liquid or gel-like environment for survival. In such an environment bacteria usually coexist with other organisms. They experience a permanent state of competition and readily establish associations of varying complexity. Strains that are entirely dependent on other organisms or that are totally independent are the exception in the bacterial world. Most strains are able to associate and keep the capacity to grow and multiply when they are alone. The most widespread type of association is the localized bacterial team, which we will examine in Chapter 4. In this chapter we will examine the metabolic functions of the bacterial cell itself.

For many years, bacteriological research centered on pure cultures of very few types of bacteria, and the resulting insights were primarily related to the cellular level of the bacterial world. It was not realized that each bacterial cell contains so few genes that it is a rather incomplete unit. Thus many activities of a pure bacterial strain might seem rather primitive, poorly controlled to the point of anarchy as a result of unrestricted growth, and very much dependent on local external conditions.

A bacterium produces a higher quantity of enzymes in proportion to its volume than does a eukaryotic cell. Some enzymes take part in the energy transport chain, some synthesize the macromolecules necessary for cell growth, and others synthesize subunits such as purine, pyrimidine, amino acids, monosaccharides, and fatty acids that are not already present in the surrounding medium.

The principal activity of bacteria, under favorable conditions, is rapid, unrestricted multiplication. Some bacteria are able to double their mass and number in approximately thirty minutes. The energy required for the synthesis of cell constituents is obtained by a much greater variety of mechanisms in bacteria than is found in eukaryotes. The synthesis of macromolecules and other components uses not only the same mechanisms as eukaryotic cells but also some additional typically bacterial pathways. Two examples of metabolic pathways that are exclusive to bacteria are the synthesis of peptidoglycan (murein or mucopeptide) and the fixation of atmospheric nitrogen (Figures 3–1 and 3–2).

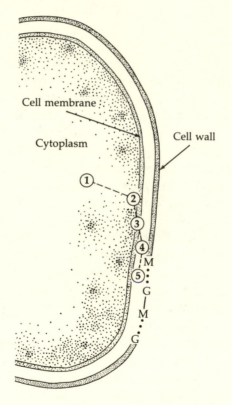

Figure 3–1. Five stages in the formation of peptidoglycan. (1) Nucleotide muramic pentapeptide is formed in the cytoplasm and (2) is transferred to a lipid carrier at the cell membrane. (3) N-acetylglucosamine, pentaglycine, and an ammonia derivative are added, probably on the inner side of the cell membrane. (4) The completed peptidoglycan unit moves through the cell membrane and links to a growing point on the cell wall (5), where peptide cross-links also form. (G = N-acetylglucosamine, M = N-acetylmuramic acid)

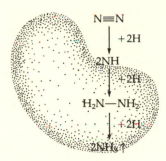

Figure 3–2. In anaerobic nitrogen fixation, nitrogen gas ($N\equiv N$) is reduced by three successive additions of pairs of hydrogen atoms. Ammonia is released as the end product. This reaction is catalyzed by the enzyme nitrogenase, containing molybdenum and several iron atoms, and requires a strong reducing agent and large amounts of ATP.

All the atoms which make up a bacterial cell must, of course, be present in its environment in order to allow its growth and multiplication and consequently to allow survival of the bacterial strain as a whole. Bacteria use hydrolyzing exoenzymes to degrade large molecules in the environment into products that are then taken into the cell. Cell permeability is usually linked to specific permeases for various substances. Using available energy sources, bacteria carry out a remarkably rapid synthesis of metabolic intermediates and normal cell constituents from small molecules actively transported into the cell.

Nutrition

The diversity of nutritional requirements found in all bacteria is far greater than that found in all eukaryotic cells. The majority of bacteria feed on organic compounds and an important minority are able to subsist on mineral (inorganic) substances. Bacteria lack the ingestive mode of nutrition, so they must take up their nutrients from solution in the water surrounding them.

From a nutritional point of view, bacteria can be divided into two groups, autotrophs and heterotrophs. Autotrophs do not require organic substances for either energy or as a source of carbon, whereas heterotrophs need organic substances as their source of carbon and energy. Autotrophs include photosynthetic bacteria that use light as a source of energy and mineral substances as their sole nutrients, including carbon dioxide as their source of carbon. Plants presently carry out the greater part of photosynthesis on Earth, while most bacteria, and to a lesser degree some fungi, ensure the conversion of wastes from all living organisms into mineral substances available to plants.

Energy-yielding Metabolism

Every bacterial cell requires energy in order to carry out its activities, particularly the synthesis of the constituents it needs for growth and cell division. In the majority of bacteria—heterotrophs—the catabolism or degradation of organic molecules constitutes the main mechanism for the liberation of energy useful to the cell. The autotrophs derive their energy either from photosynthesis (which is carried out by certain bacteria and by algae and plants) or from chemolithotrophy, inorganic chemical reactions such as the oxidation of ammonia, hydrogen, or methane. Chemolithotrophy is only known in bacteria, but both bacteria and eukaryotes link energy release with biosynthesis through chemical reactions involving adenosine triphosphate (ATP) and pyridine nucleotides (NADP, NAD).

Bacteria possess a much greater variety of energy-yielding mechanisms than eukaryotes. Eukaryotes use only a very small proportion of the prokaryotic mechanisms and do not possess any exclusively eukaryotic energy-yielding mechanism. A single bacterium can resort to two or three mechanisms and generally chooses the one which produces the optimal energy for its given circumstances. Bacteria are unable to benefit directly from the energy released from the catabolism of naturally-occurring nutrients of high molecular weight outside the cell because bacterial membranes do not normally permit the entry of these nutrients into the cell. Instead, exoenzymes outside the cell break these substances down into smaller permeable substances, releasing energy that cannot be utilized by the cell. The principal mechanisms in bacteria for the utilization of energy sources are photosynthesis, fermentation, and respiration.

Photosynthesis. This complex mechanism is responsible for the conversion of light energy into chemical energy and the subsequent conversion of carbon dioxide into organic substances (Figure 3–3) It is found in a few particular groups of prokaryotes living in sulfur environments where their usual competitors, plants and eukaryotic algae, cannot prosper, as for example in sulfur-rich muds or well-lit hot water. Most photosynthesis on Earth is carried out by eukaryotes (algae and plants) by means of their green plastids. These chloroplasts closely resemble cyanobacteria, previously called blue-green algae (54), and it is likely that ancestral cyanobacteria gave rise to chloroplasts

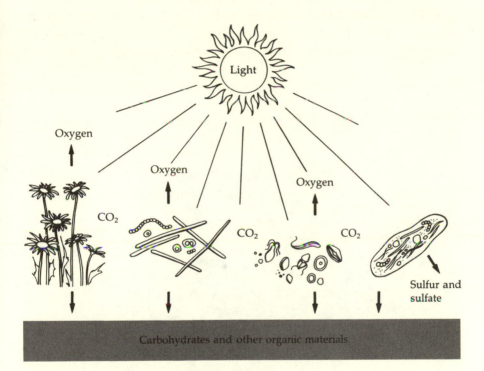

Figure 3–3. In the presence of the pigment chlorophyll, the complex reactions of photosynthesis transform light energy from the sun into chemical energy. The chemical energy can then be used to reduce carbon dioxide to organic compounds that can be used by other organisms. Green plants, algae, and cyanobacteria use water to obtain reducing power and give off oxygen as a by-product of their photosynthesis. Anaerobic purple and green bacteria receive their reducing power from substances in their environments such as hydrogen, hydrogen sulfide and other sulfur compounds, and some organic compounds.

(32,35,36,55). In prokaryotes, photosynthesis is carried out in the presence of free oxygen (aerobiosis) by the cyanobacteria (Figure 3–4). Under anaerobic conditions, photosynthesis is carried out by cyanobacteria or by green or purple bacteria. In all living organisms, the mechanism of photosynthesis involves pigments (chlorophylls and carotenoid pigments), lipids, proteins, and, of course, electron carriers.

Cyanobacteria, like plants, possess β-carotene and chlorophyll a. They live at the surface of bodies of water and release free oxygen as a result of their type of photosynthesis. Their ancestors are most likely to have been responsible for the first appearance of oxygen in the terrestrial atmosphere, leading, later, to the advent of aerobic bacteria and, much later, to the appearance of eukary-

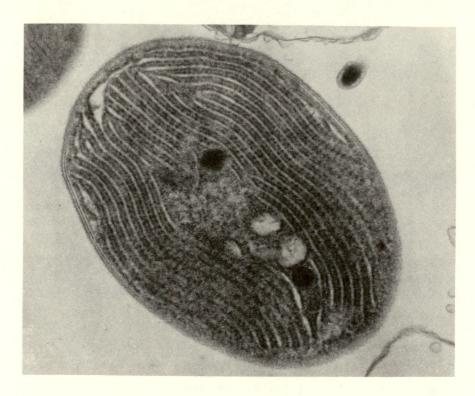

Figure 3–4. Micrograph of a cyanobacterium, in which the photosynthetic membranes are visible as parallel stripes. (*Aphanothece*; ×41,250. Courtesy of John Stolz, Boston University.)

otes (33,35,36). Certain types of cyanobacteria are able to fix atmospheric nitrogen. These are among the living organisms with the simplest nutritive requirements: carbon dioxide and nitrogen from the air, water, and mineral salts.

Green and purple bacteria grow in specific anaerobic environments where they absorb wavelengths of solar radiation not used by aerobic photosynthetic organisms. They use bacteriochlorophylls and carotenoid pigments for their photosynthesis. Their hydrogen sources include hydrogen sulfide or organic molecules in their habitats in the mud of shallow swamps, in intermediate zones of deep lakes with no major currents, or in marine tidal flats. Some fix nitrogen and may release free hydrogen as a final product of catabolism. These photosynthetic bacteria appear to be vestiges of ancestors that played an important role before cyanobacteria appeared and oxidized water to produce free oxygen.

Fermentation. In fermentation, energy results from complex oxidation–reduction reactions based on transfers of electrons that originate from a larger organic donor and finally combine with an acceptor that is also an organic molecule. The donor is usually a sugar, but in bacteria it may also be an amino acid, another organic acid, or the base portion of a nucleotide. A great variety of acceptors yielding fermentation waste products of all kinds are found in bacteria. The waste products include ethanol, methanol, lactic acid, butyric acid, pyruvic acid, acetic acid, and many others. Fermentation occurs under anaerobic conditions and constitutes the sole source of energy for several kinds of strictly anaerobic bacteria (Figure 3–5).

Respiration. A series of complex oxidation–reduction reactions also yields energy from respiration. In respiration, the hydrogen donor is generally an organic molecule and the final hydrogen (electron) acceptor is an inorganic substance. The highly complex electron transport chains involved in bacterial respiration are located in the cytoplasmic membrane and its intricate folds, the mesosomes (Figure 2–3). Respiration yields more energy than does fermentation. Nearly all naturally-occurring organic substances of low molecular weight can be used as electron donors by certain bacteria. In anaerobic respiration, the acceptors are substances other than oxygen, generally nitrates or sulphates. Bacte-

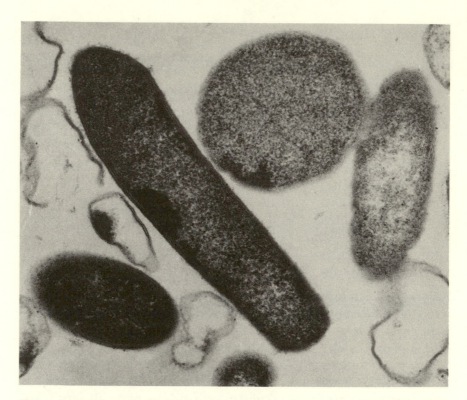

Figure 3–5. Micrograph of a fermenting bacterium, *Clostridium Ramosum.* (× 50,200. Courtesy of David Chase, Veterans Administration Hospital, Sepulveda, California.)

ria that carry out anaerobic respiration may be strict anaerobes, facultative aerobes, may utilize oxygen in lower than atmospheric concentrations (microaerophylls), or may tolerate it without using it. In aerobic respiration, the final electron acceptor is free oxygen as in eukaryotes. Aerobic respiration produces the highest yield of energy useful to the cell. This explains why most bacteria are more effective in aerated soils and also why they carry out the final stage of water purification more efficiently in the presence of oxygen. Bacteriological techniques for the study of aerobic bacteria require large surface areas for contact with air. In industrial microbiology, sterile oxygen is percolated through the containers used for the cultivation of aerobic bacteria in order to maximize the yield.

Coordination and Regulation

General considerations. Regulation of prokaryotic activity occurs at many levels: transcriptional, translational, metabolic, and behavioral. The extent to which it also occurs at the population and community levels has been seriously underrated. The great majority of biologists, influenced by the fact that bacteria are unicellular and, for the most part, endowed with great autonomy, believe that in bacteria all phenomena of coordination and regulation occur only at the cellular level. They readily accept the notion of social animals, particularly social insects, but have not yet considered the concept of social or cooperating microorganisms.

In the bacterial cell, excellent coordination can be observed between those metabolic pathways involved in the acquisition and transport of energy and those that carry out the biosynthesis of various necessary substances. This coordination is influenced by the external concentration of nutrients in the cell's immediate environment.

The type of regulation that is most rapid and most sensitive to immediate needs in a bacterial cell is accomplished through a modification of enzyme activity. The best known example of this process is the inhibition of the activity of many enzymes as a result of the accumulation of the end product of their respective metabolic pathways. As soon as a cell exhausts the supply of an inhibiting end product, the previously inhibited enzyme resumes its activity. This mechanism is not only very rapid and extremely responsive, but is also completely reversible.

The other important regulation is realized by the inhibition of enzyme synthesis and occurs at the transcriptional level. It is carried out by means of regulatory genes which control nearby genes coding for the enzymes of a given metabolic path (25). We have seen that such a group of biosynthetic genes and its respective regulatory genes in most cases form the unit called an operon, a series of genes whose transcription can be blocked by a specific repressor. The repressor is a protein, synthesized by a regulator gene, that represses the series of genes when the end product of the metabolic pathway is present. Because the active lifespan of bacterial messenger RNA is barely a few minutes, the quantity of enzyme synthesized is directly dependent on the number of messenger RNA molecules that have been transcribed. In the phenomenon of enzyme induction, a carbon substrate, present in the immediate environment of the bacterial cell, generally derepresses the gene that codes for the enzyme initiating the catabolism of that substance. When several substances outside the cell serve as energy sources, cellular coordination mechanisms generally enable the cell to catabolize the most favorable source exclusively, and to turn to another slightly less favorable source only when the most favorable source has been exhausted. Regulation at the transcriptional level is not as rapid as regulation involving a modification of the activity of an already synthesized enzyme. Once an enzyme has been synthesized, it remains present in the cell even after its synthesis has been stopped.

These two regulation mechanisms complement each other. For example, an excess of the end product of a metabolic pathway involving a series of enzymes that are coded for by an operon will inhibit the synthesis of all the enzymes in the series, in addition to inhibiting the activity of the first enzyme of the pathway.

The synthesis of stable RNA molecules (ribosomal and transfer RNA) is also carried out in an orderly fashion, coordinated with the availability of nutrients and with cell growth. The mechanisms involved in this coordination are only partially known. Similarly, very little is known about the excellent synchronization of the synthesis of DNA of the genophore with the sequences of bacterial cell division. However, rare instances have been observed where the imperfect coordination of these two processes produces unbalanced growth.

Cell wall synthesis and cell division. Another example of coordination is found in the mechanism of cell wall synthesis. This mechanism ensures that cell growth and division into two offspring cells takes place without the formation of holes in the cell wall during the process of replacement of its elements. The cell wall is expanded and its form is modified during growth, but during this change it still protects bacterial cells from osmotic lysis in their general environment. The other components, including DNA, different RNA molecules, proteins, amino acids, and sugars, are carefully multiplied by two and the formation of the two identical copies of the cell, which eventually separate, is conducted with precision, under genetic regulation. The rapidity of population growth is such that any bacterial strain may lose a certain proportion of its cells by death without jeopardizing its future survival. Bacteriostatic factors suspend DNA replication as well as all other bacterial activities, including the release of enzymes. For example, the decomposition of food can be retarded by refrigeration and, even more successfully, by freezing.

CHAPTER 4

Bacterial Associations

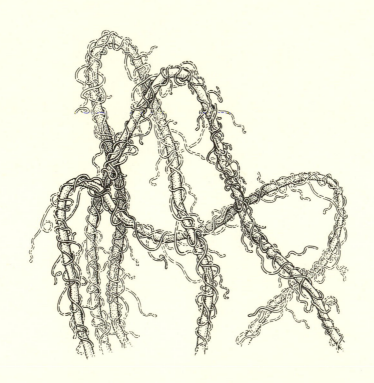

BACTERIA ARE found everywhere over our planet's surface, in the natural waters, in soil, and in the digestive tracts of all animals. In most cases they interreact with their living neighbors, competing or collaborating with them. An uninhibited multiplication in normal natural conditions provokes fierce competition among the different bacterial cells present in the same place. The fittest cells—those most capable of leaving many offspring under the specific conditions—are selected constantly. Thus, the Earth's bacterial population is constantly rejuvenated and kept in the best possible condition. The high degree of specialization in bacteria permits their successful development in conditions of competition and of cooperation. This extreme specialization arises from the type of prokaryotic differentiation discussed earlier, where the quantity of intracellular genes of each strain is kept at a minimum, and just the genes required for survival and replication are retained, mostly on the genophore. If necessary, some intracellular genes can be replaced, since every bacterial cell has potential access to the genes of other strains through mechanisms of exchange. Permanent selection then ensures the survival of those cells that have made the best adaptation. The same mechanisms help bacteria to adjust to cooperation. Their associations with other living things are frequent and everywhere present. It seems to be a natural way of life, as if bacteria, which possess limited genetic information in any cell and therefore have very specialized and rather incomplete metabolic possibilities for one single cell, can benefit by combining their genetic and metabolic possibilities with those of their neighbors.

All bacteria possess receptors for phages; many also possess receptors for foreign soluble DNA. An estimated half of all bacterial strains, including practically all Gram-negative ones, will accept conjugation plasmids (17). Each bacterial cell is thus receptive to genes from different strains. Every bacterium continuously disseminates many of its own genes, often packaged as shared small replicons, for the possible benefit of different strains. The mechanisms of transformation, transfection, lysogenization, transduction, and conjugation enable each bacterium to be both a potential emitter and a receiver of these information molecules, the genes. The very high proportion of energy that is used up for the operation of this communication network may seem surprising in view of the fact that a bacterium possesses a very limited intracellular genome, reduced to what is essential for its super-specialized life. Moreover, the probability that a single bacterium may benefit from the random exchange of genes occurring in nature is infinitely low, billions of times lower than our odds of winning when we buy a lottery ticket. Thus, most of the Earth's bacteria will not benefit for hundreds of years from this social investment in systems for exchanging information with other strains. However, this important altruistic activity, continuously regulated by different selective pressures, is the basis for the totally unique rules of the bacterial way of life. Bacteria have set up a tremendous gene-sharing mechanism that manifests its exceptional potential at the broader levels of the bacterial world, and involves an important geographic fraction of all of Earth's prokaryotes.

Instead of forming associations with other bacteria, bacterial strains sometimes depend on eukaryotes, living in symbiosis with animals, plants, green algae and other protists, or fungi. It seems evident to us that bacteria exhibit four levels of organization and function: individual cells with their strains that sometimes form colonies in nature; localized teams with many different complementary strains, showing varying degrees of symbiosis; the planetary entity of all prokaryotes, in which the shared visiting genes play a major role; and the association of prokaryotes with eukaryotes. In any of these associations, from permanent symbiotic relationship to very temporary cooperation, the bacteria involved can easily grow to provide the numbers required and the precise type of complementary metabolism needed. The resulting utilitarian and reversible associated entity obtains a higher yield from the available nutrients, for a longer period and within a buffered localized zone, stabilizing the associated cells of different types. In this chapter we will discuss cooperation between different bacteria at the levels of the localized team, the planetary entity, and associations with eukaryotes.

Localized Bacterial Teams

Most bacteria, even when mixed in cooperating teams, exist as isolated cells. In the laboratory or in fast-moving water they do not adhere to other bacteria to form structures or morphological entities visible either to the naked eye or through the light microscope. However, scanning electron microscopy and certain other specialized techniques have clearly shown that in many cases bacteria adhere to surfaces in an ordered configuration, an arrangement particularly useful in preventing them from being carried away by liquid currents. This is the case with bacteria coating rocks and plants in the beds of waterways, with healthy intestinal bacteria, and with dental plaque bacteria. In sediment, on the surfaces of other organisms, and on lake and ocean surfaces bacteria may form visible structures. The important bacterial teams in soil and in rumen, those involved in decomposing animal cadavers, and those purifying water do not exhibit fixed visible configurations (see Figure 2–2). However, sometimes they have a structure like a mat, as in Laguna Figuroa (see cover illustration). The discovery of teams composed of different strains of bacteria came about through the study of their complementary enzymatic properties. This reciprocal activity allows certain bacterial strains to act simultaneously. Others work sequentially in a cascading succession of activities, where each type of metabolic strain takes up the work at the point where the preceding strain has left off. Here the particular action of each strain is easily observed, since the bacterial composition of the team changes with each successive phase of activity. The principal phenomenon that enables these teams to carry out activities greater than the sum of the capacities of each constituent strain is the complementarity of their enzymes (22,24).

Surprisingly, strong competition within these groups is ultimately responsible for their remarkable stability. Once the most appropriate cells are selected for each sector of the team, it is difficult for another, less suitable, cell to usurp them. The community of bacterial cells constituting different teams is always open to new members, but selection is based on immediate merit, and favors the most suitable strains, generally those that have already undergone a lengthy evolution as members of the team.

Mechanisms responsible for the success of bacterial teams. In a task carried out by a bacterial team, such as the decomposition of accumulated organic matter, the necessary stages are carried out with great economy as a result of the cumulative effect of complementary enzymes contributed by different strains. Sometimes these enzymes are released simultaneously, but more often they are released in succession. The strains responsible for several stages in a task will continue to multiply for a longer period of time than those strains that play a smaller role; consequently, they will outnumber the latter when the task is completed.

There is great variation in the size of bacterial teams. However, even the smallest ones, those living in the digestive tubes of tiny worms or insects, are very complex. Numerous studies have demonstrated that soil, particularly a rich arable soil, is host to hundreds of different bacterial strains, as well as to eukaryotic microorganisms, fungi, and minuscule plants and animals (24). Some of these studies have also shown that these individual organisms together form a more complex, true living organism (22). An extensive division of labor exists among bacterial teams where two or three types of strains generally play similar roles, thus maintaining a healthy competition between them and facilitating the acquisition of extrinsic genes, if necessary (23).

The well-known bacterial teams active in soils, in the rumen of animals, and elsewhere help to maintain a stable environment (pH, trace gas composition, and so on), and their specific activity stems from the reciprocal benefit derived by each strain from its association with the others. Community activity in these teams generally takes the form of enzyme associations. In this way their enzymes complete each other, enabling the environment to support a higher bacterial population for longer periods than would otherwise be possible. When the favorable substrates occur irregularly, the longer, more efficient metabolic activity of the association of strains maintains its multiplication for longer periods than would be expected in the case of a single strain. The more intense utilization of substrates by teams of bacteria results in higher concentrations of cells, of enzymes, and of genes for potential exchange. Teams composed of several types of bacteria carrying out a true division of labor benefit from the regulatory systems of the bacteria involved. A great variety of enzymes is provided by the mixture of cell types, and several mechanisms of intracellular coordination in neighboring cells are affected by nutrients in the common environment. Selection processes are favored, giving rise to the appearance of intermediate forms of bacteria exploring new, marginal metabolic possibilities. The bacterial team creates its own microclimate, where sometimes even the temperature is stabilized at an optimum level. The end product of the catabolic chain of one cell constitutes the source of energy for another strain, and thus its release contributes to the regulation of the multiplication of that strain. This interdependence decreases each strain's autonomy but is beneficial to the entire association (54).

In processes involving a succession of activities, such as corpse decomposition, an alternation of bacterial forms can be observed in addition to the synergetic action of the strains involved. Molecules released by one strain stimulate the multiplication of the next strain in the succession. This strain is, in turn, largely replaced by another, once its preferred substrate is exhausted. This directed selection is another mechanism that, in large bacterial teams, determines both the types of bacteria present and their numbers. In the case of balanced teams, such as those in herbivore rumen and in soils, this mechanism ensures the general stability of the over-all group of constituent bacteria despite cyclical or seasonal variations. The presence of several sporulating strains favors the alternation of strains in the teams; these strains are active as long as their nutrients are plentiful, but become totally inactive during periods of scarcity and survive only in the form of spores.

The other social activity in localized bacterial teams is the exchange of genes. The probability that a bacterium in a team may obtain a favorable gene is still very low, although it is greater than for isolated bacterial cells or strains. Nevertheless, the mass of bacteria involved and the variety of strains present do significantly increase the possibility that some new visiting shared replicons or a few isolated genes might be accepted and multiplied. If these genes confer even a slight advantage to the host cells, this amplifying activity will be tremendous. The presence in bacterial teams of a concentrated population of many different strains favors genetic exchange in that some strains can serve as stepping stones in the transfers of genes between strains that are metabolically distant (42). In addition, different teams may exchange genes if they are not too far from each other's reach. For instance, our intestinal bacterial teams exchange cells and genes with the bacterial teams of the soil surrounding us (17).

It is probable that genetic exchange is also enhanced in highly concentrated mixtures of different bacteria. However, at the level of the functions of bacterial teams, the division of labor due to the complementarity of enzymes appears to be much more important than genetic exchanges. It is easy to see that even very large teams possess only an infinitesimal fraction of the prodigious variety of all bacterial genes. Although the genes belonging to the entire team become more accessible to each member due to their physical proximity, their variety, nevertheless, remains limited.

Examples of the activity of bacterial teams

Rumen. The rumen, first of four stomach cavities of ruminants (cud-chewing mammals), and of bovines in particular, is the site of the activity of abundant and diversified microbial populations, for whom it constitutes a particularly favorable ecological niche (Figure 4–1). From a biological point of view, the rumen is an example of ectosymbiosis (a reciprocal, often beneficial, association between two or more different organisms in which one organism does not penetrate the other's cells or tissues) of different bacterial strains between themselves and with a ruminant. Countless instances of bacterial ectosymbiosis and endosymbiosis (a symbiosis involving penetration of cells or tissues by one member of the association) with insects, birds, and mammals occur in nature. The rumen and its contents form a microbial community that is of great interest. Processes that take place in such an environment are comparable to those increasingly used in industrial bacteriology in systems of continuous bacterial culture.

The rumen receives a mixture of incompletely chewed food and saliva. The microbial population of the rumen then acts upon this mixture for several hours. Thus transformed, the content of

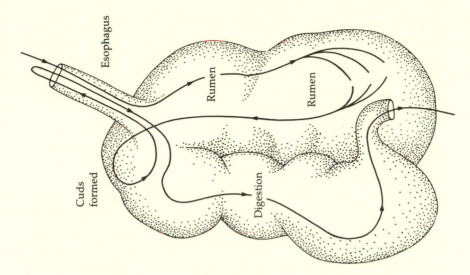

Figure 4–1. Microbial fermentation occurs in the rumen. The food mass is then formed into cuds, regurgitated and chewed again, and digested.

the rumen continues its journey into the other gastric compartments and the intestine. The foods entering the rumen are rich in cellulose and poor in proteins and lipids. Saliva does not contain cellulase to digest the cellulose. However, the rumen's immense population, billions of microbes per milliliter, includes specialized bacteria exercising cellulolytic action, resulting in the production of glucose and cellobiose. These substances, after additional microbial action, can then be directly assimilated by the host.

The bacterial microbiota of the rumen consists of strict anaerobes, both Gram-positive cocci and Gram-negative rods. Although the strains involved and their biochemical capacities can be identified relatively easily, the action they may exert on one another is much more difficult to analyze. The rumen also contains a large population of ciliates that feed on bacteria and their metabolites. Their existence has been recognized since 1843, but their study has been difficult because of bacterial contamination.

The contents of the rumen and its localized environment remain relatively constant. In bovines, a copious supply of saliva facilitates enzymatic reactions and acts as a buffer in maintaining the pH (5.8 to 6.8) of the hundred or so liters of semiliquid mixture. The bovine rumen is maintained at a temperature of 38.5 to 39°C and at a redox potential favorable to the multiplication of anaerobes. Bacterial and ciliate constituents are digested by proteolytic enzymes in the lower stomachs and intestinal segments, producing ammonia, urea, and amino acids. These substances are then reassimilated by the ruminants. The vitamins produced here are also mostly of bacterial origin. Thus, the rumen can be compared to an industrial fermentor, in which a substrate is subjected to the activities of the appropriate bacterial associations, ensuring the continuous production of useful metabolites (22,24).

Soil. The soil is an underestimated living entity. The role of soil in the perpetuation of life on this planet has been recognized since our earliest ancestors first began to practice agriculture. They noted the differences in fertility of various soils and observed the formation of humus from wastes and plant and animal corpses. They were able to conserve or restore this fertility, which benefited plant life, by the practice of fallowing, that is, allowing the soil to rest, or by the addition of organic materials called compost. In short, the observation and empirical exploitation of the biological phenomena occurring in the soil long preceded the recognition of their causes.

The identification and the experimental reproduction, inhibition, and restoration of biological activities of the soil also preceded scientific understanding by approximately ten years. Interdisciplinary studies demonstrated the chemical nature of these reactions, their inhibition by sufficiently long and intense heating, and the restoration of the initial properties of inactivated soil samples by mixing in a small quantity of untreated, normal soil. These studies demonstrated the biological nature of these well-known phenomena long before the discovery of their causes.

The late advent of soil microbiology

The misunderstanding of the microbiological nature of soil phenomena was ended by a series of discoveries that demonstrated the role of microorganisms in producing and conserving soil fertility and thus in maintaining both plant and animal life on dry land. The new understanding arose in large from the work of

S. Winogradsky and M. W. Beijerinck, who were active in Europe early in this century, twenty years after Pasteur's discovery of the microbial origins of infectious disease in man and animals. As we mentioned earlier in this book, bacteriology was long dominated by the interests of medical and veterinary groups, rather than by the more general biological considerations related to the microbiological phenomena of soils. However, in 1949 P. Goret and L. Joubert, two French specialists in veterinary microbiology, described the soil as a misunderstood living entity (22). They may be credited with having predicted more than thirty years ago, long before the development of bacterial genetics, the importance of certain group activities of bacteria and certain facts that are now the object of our attention.

Abundance of the bacterial population of soils

Statistics regarding the bacterial content of soils amply demonstrate the magnitude of the very large population of soil bacteria. There are from 1 to 10 billion bacterial cells per gram of soil, and from 1700 to 3600 kg of bacteria per hectare. The latter figure compares with 1700 kg per hectare for fungi and 170 for protists. These are rough estimates, since many potential sources of error, such as whether the sample is representative, and what counting methods and techniques are used, influence these figures. For example, microscopic examination reveals approximately 100 times more bacteria in a given volume of soil than do culture counting techniques that reflect the number of cells able to multiply on artificial media. This disparity can be attributed to the highly specific and very diverse nutritive requirements of soil microorganisms. All researchers agree that at present it is impossible to carry out precise measurements of the numbers of living microorganisms in soil. They also agree on the biological consequences of the presence of such important masses of bacteria in the soil. Figure 4–2 shows the large reservoir of bacteria in the soil and at the bottom of bodies of water. Soil and the contents of the digestive tracts of animals contain more than one billion bacteria per milliliter, and sludge at the bottom of bodies of water contains almost as many.

Figure 4–2. Major bacterial concentrations in nature.

*Principal strains of soil bacteria and
their functional variety*

The principal types of soil bacteria that can be grown in pure culture are *Actinobacteria, Bacillus, Clostridium, Flavobacterium,* and *Pseudomonas.* The soil maintains its over-all fertility despite the constant removal of nutrients by all plants. This stability is due to the recycling of nutrients by plants and animals, either from waste products of their vital activities (leaves, excrement) or from constituents of their corpses. However, most of these substances cannot be directly assimilated by plants, which can only absorb mineral substances. Soil microorganisms are responsible for the transformation of constituents from dead animals and plants into mineral molecules.

Both the organic and inorganic constituents in the soil are subjected to the action of bacteria of very diverse metabolic capacities. The relative proportions of the microorganisms involved in these processes are dependent on the physico–chemical properties of their environment (chemical composition, aeration, compactness, pH, humidity, and temperature). These bacteria can be autotrophs or—more often—heterotrophs; aerobes or anaerobes; mesophiles, thermophiles, or psychrophiles.

In addition, certain bacteria carry out specialized activities such as the degradation of cellulose, lignin, or pectin; the synthesis of organic nitrogen-containing molecules by fixation of atmospheric nitrogen; and the oxidation of ammonia to nitrites or of nitrites to nitrates. These natural phenomena were first exploited empirically and later scientifically, in agriculture in the practice of soil enrichment, and in sanitary engineering in the purification of used water.

The general activity of soil bacteria can be analyzed by an examination of the principal characteristics of the cycles of living matter which occur in the soil. The two most important cycles are the carbon cycle and the nitrogen cycle.

Organic compounds containing carbon were initially synthesized from carbonates present in fresh and salt water as well as in soil. Now, however, atmospheric carbon dioxide constitutes the primary substrate for the chlorophyll-based synthesis carried out by plants in the presence of sunlight. Although photosynthetic bacteria originally played the key role in the synthesis of organic compounds, their present role is relatively minor.

Soil bacteria degrade organic molecules, derived from plant and animal tissues that have returned to the soil, into inorganic products (Figure 4–3). Compounds from plant tissues are more abundant than those from animal tissues and are more difficult to break down, particularly cellulose and ligneous substances. Either aerobic or anaerobic bacteria, depending on environmental conditions, carry out this catabolic process. The oxidation of plant tissues is incomplete, since lignin, tannins, fulvic acid, kerogen, and waxes resist the oxidizing action of microorganisms for a long time. Thus, humus—a brownish, colloidal residual organic product that undergoes a very slow degradation—is formed. This complex substance confers to the soil its texture, hydrophilic capacity, and resistance to erosion. The interruption of humus degradation in humid anaerobic conditions leads to the formation of peat. The end product of humus degradation is kerogen.

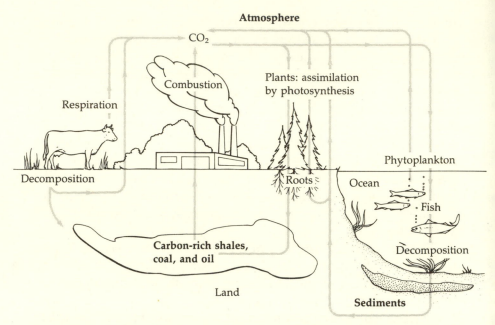

Figure 4–3. Carbon dioxide is removed from the atmosphere by photosynthesis in plants and other photosynthetic organisms. Respiration and decay and combustion of fossil fuels return carbon dioxide to the atmosphere. In anaerobic conditions over enormous spans of time, products of decay containing carbon became part of fossil fuels.

Nitrogen is essential for protein synthesis in plants and animals. Bacteria are not only active in the decomposition of nitrogen constituents of plants and animals for their subsequent transformation into mineral molecules, but they are also the only living cells active in the synthesis of organic nitrogen-containing compounds through the fixation of atmospheric nitrogen (Figure 4–4). There are two types of nitrogen fixation. In symbiotic fixation, bacteria of the *Rhizobium* type penetrate the roots of leguminous plants where they form nodules. Another type of nitrogen fixation occurs in conditions prevalent in tropical regions, where various bacteria such as *Azotobacter*, *Clostridium*, and cyanobacteria are able to fix nitrogen without plant symbiosis.

The degradation of the nitrogenous constituents of plant and animal tissues and of the waste products of their vital activities forms the most important aspect of putrefaction phenomena in animals. These processes are essentially the same for the degrada-

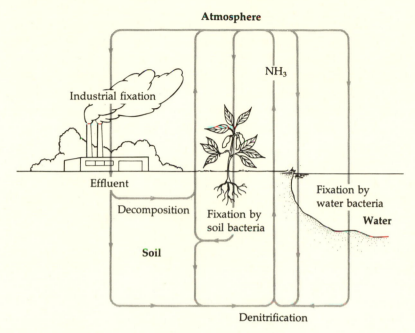

Atmosphere

NH₃

Industrial fixation

Effluent

Decomposition

Fixation by soil bacteria

Fixation by water bacteria

Water

Soil

Denitrification

Figure 4–4. Nitrogen-fixing bacteria are the primary agents that transform atmospheric nitrogen to nitrites and nitrates that can be used by other organisms. Denitrifying bacteria return nitrogen to the atmosphere. Industrial nitrogen fixation involves relatively little nitrogen compared to the amounts fixed by bacteria.

tion of nitrogenous compounds of dead plant tissues. They are also involved in the transformation of animal excretory products such as urea and uric acid. All of these substances are transformed into ammonia by functional teams of bacteria with extensive catabolic capacities. The resulting ammonia or ammonium salts are subsequently converted by other bacteria to nitrates assimilable by plants.

All of the diverse functional types of bacteria that we have mentioned coexist in the soil. We have seen that their proportions, their individual and complementary activities, and their effects are all influenced by factors intrinsic to the populations themselves and to their environmental conditions. Phenomena such as biological competition, antagonisms in certain cases, and favorable associations of biotypes also take place simultaneously. The presence of truly antibiotic strains in the soil was first predicted by Pasteur. This prediction was later unequivocally demonstrated by the work of S. A. Waksman in the U.S.A., René J. Dubos in France and in the U.S.A., and others, who showed that certain *Streptomyces* and certain eubacteria produce highly active antibacterial substances such as streptomycin, tyrothricine, chloramphenicol (chloromycetin), and neomycin. It is interesting that some of these discoveries, which have had numerous important applications, were derived from research that initially had no practical goal.

Bacteria of natural waters. Just as bacteriology primarily centered on the study of infectious agents in man and animals, early research on the bacteria of natural waters focused mainly on the detection of pathogenic bacteria in aquatic environments. As had been the case for soil bacteriology, both practical and technical reasons delayed the systematic study of the bacteriology of water. For many years the search for pathogenic bacteria and indicator bacteria signaling the potential presence of pathogenic bacteria held the center of attention.

Even in the absence of extraneous contaminations the bacterial population of natural waters is varied and abundant. Its composition and density depend on the properties of the water in question, such as its salt content, flow rate, degree of oxygenation, and concentration of organic substances. Much of the bacterial population of natural waters is of soil origin and can be enriched by contributions from the air and from effluents, produced by animals or human activities, that may contain organic wastes and various chemical compounds.

Self-purification of fresh water

Water polluted by the addition of residual organic matter of domestic, agricultural, or industrial origin is normally subject to self-purification by means of a degradation of organic matter similar to that undergone by plant and animal residues in soil. The self-purification of water is the result of ecological cycles in which bacteria, algae, and other eukaryotes participate. Bacteria are the initial and final agents of these successive changes and intervene as functional bacterial teams of the natural water microbiota. Their numbers and types as well as the succession of phenomena that they carry out are determined by various factors that can enhance, retard, or even totally inhibit self-purification.

The principal factor controlling self-purification is the abundance of the organic pollutants. The quantity of pollutants determines the availability of dissolved oxygen in the water. The oxygen content is measured as the biochemical oxygen demand (BOD) index. The depletion of dissolved oxygen caused by the intensity of bacterial phenomena, and the presence of pollutants, can be more or less compensated for by contributions of atmospheric oxygen. These contributions are very small where the water surface is dormant, as in ponds or very slow-moving streams, and more significant if the surface is agitated, as in swiftly-running rivers and waterfalls. Where a body of water is reoxygenated by agitation, self-purification takes place rapidly and completely through the intervention of aerobic bacteria; otherwise, it is carried out less efficiently by anaerobes.

The analysis of the distribution of bacterial biotypes participating in the self-purification of water and their respective roles is as difficult as the study of bacterial teams in soils. Although it is well known that bacteria play the principal role in this process, the participation of other microorganisms such as algae and other protists, and fungi, must also be considered. Many protists, which are ingested by small animals, are efficient predators of bacteria. Algae play an important role in the reoxygenation of water by photosynthesis, and the periodic excessive multiplication of algae is a phenomenon that is, in turn, dependent on substances degraded by bacteria.

Treatment procedures for used water

In developed regions of the world, used water is treated before it is released into a natural body of water. Treatment procedures are applied in order to eliminate abundant organic substances, and particularly pathogenic microorganisms, as much as possible, as well as to prevent direct discharge of industrial wastes, especially toxic substances, into waterways.

Water treatment methods essentially recreate under favorable artificial conditions the processes of natural self-purification. The specific techniques involved depend on the nature of the effluents to be treated. The principal sources of water requiring purification include industrial used water containing organic effluents such as those from pulp and paper plants and food industries, used water of agricultural origin (principally from industrial husbandry), and, particularly, used water of domestic origin.

The choice of method, the sequence of processes, and the evaluation of their respective results are based on the results of microbiological testing and BOD measurement. The principal methods are the trickling filter method, the use of activated sludge, and the action of anaerobic digestion in a septic tank.

The trickling filter method primarily exploits aerobic processes (Figure 4–5). Used water is finely sprayed onto a bed of particles arranged in increasing size from the surface to the bottom. Within several weeks, the surface of this bed becomes covered with a viscous microbial slime, mainly composed of bacteria. The superficial layers of slime also contain fungi, which play a limited role, and algae, whose proliferation may interfere with the normal course of filtration. Protists, insects, and arachnids are also present, with a minimal participation. Filtration through this bed lasts from twenty to sixty minutes. During this process, the filtrate is almost completely cleansed of organic matter, either by precipitation or by degradation carried out by bacteria. A second

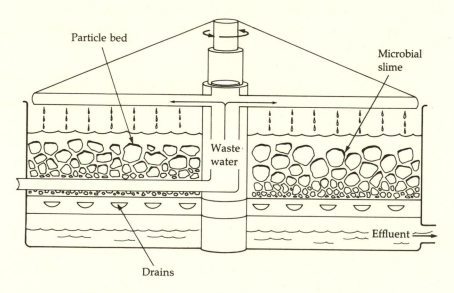

Figure 4–5. The trickling filter method of used water treatment.

sedimentation filters out any remaining precipitable microorganisms and produces a clear effluent containing very little organic matter.

Treatment of used water by activated sludge calls for the intervention of a true functional bacterial team that forms a flocculating (lump-forming) oxidative slime out of the sewage. After resting in a succession of closed tanks, the sediments are treated so as to both reduce their volume and destroy pathogenic microorganisms (Figure 4–6). Protists play an important role in complementing bacteria in this method. Many of them feed on bacteria and provide food for very small animals.

The action of anaerobic digestion is utilized in domestic septic tanks (Figure 4–7). It is preceded by the sedimentation of substances in suspension that are then subjected to a self-sustaining anaerobic degradation, releasing hydrogen sulfide. The effluent still shows a high BOD and is only partly purified. Anaerobic digestion can be adapted to industrial, agricultural, or municipal requirements. The process is carried out in large enclosed tanks or digesters in which anaerobic bacteria attack organic matter and release various gases such as hydrogen sulfide, hydrogen, nitrogen, carbon dioxide, and methane. The resulting

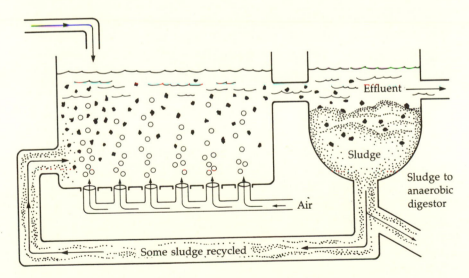

Figure 4–6. Activated sludge treatment for used water.

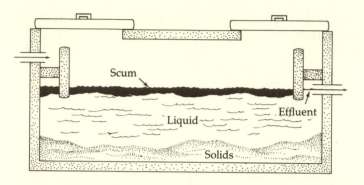

Figure 4–7. Anaerobic digestion in a septic tank. Soil bacteria in the leaching field complete the degradation of wastes in the effluent.

sludge, after drying, is a convenient soil fertilizer. Combustible gases released in this process have found use as energy sources, although they are rather limited in quantity. The digester receives fresh sludge combined with sludge from a previous digestion. The bacterial population active in anaerobic digestion is primarily composed of methanogenic anaerobes (5,59).

Bacterial Cooperation at the Continental or Planetary Level

The concept that there exist functions of a complex type involving the bacterial population of one or more countries, a continent, or our whole planet is just beginning to be accepted in biology.

> *Facilities for the exchange of information and communication molecules among all bacteria, combined with favorable selection pressures*

These functions are implemented by relatively simple means, namely by communication and the exchange of information, natural selection, and distribution of information by competition and differential growth. With many ways of exchanging genes (information molecules), each bacterial cell has the capacity to send and receive genes, and can, therefore, be compared to a two-way radio. Thus, these genes act not only as transmitters of hereditary traits to offspring cells at the time of reproduction, but also act all the time as molecules of information and communication among different bacteria.

These methods of reciprocal exchange are analogous to the very complex man-made networks for receiving and transmitting information at a distance. In bacteria, periods of frenetic exchange activity are interspersed with lengthy periods of quiescence characterizing phases of stability. Maximal genetic exchange occurs at the onset of new unfavorable or potentially more favorable circumstances challenging many bacteria, realizing a general selective pressure.

A planetary bacterial entity—a true superorganism

We are convinced that bacteria constitute a community of organisms, invisible to the naked eye, that is extremely varied, dispersed to the point of being everywhere on the surface of the Earth, and very efficient in critical situations. Due to their common potential genome and common origin, it is possible, in fact, to speak of all bacteria as constituting a true superorganism.

We believe that when environmental conditions exert selective pressures on bacteria in a large area, or perhaps even across the entire planet, the exchange of genes, normally entirely random, proceeds from the few cells possessing the favorable gene to all strains requiring it. A geographical diffusion in all directions takes place, aided by wind, water currents, and migratory animals. In terms of physiological distance as a reflection of the degree of metabolic difference between participating strains, the transmission of favorable genes tends to be much more rapid between strains whose metabolisms are closely related.

Like an electronic communications network, the bacterial world possesses an enormous data base, in this case in the form of bacterial genes. Both also possess a mechanism that permits them to choose the right solution to a specific problem. In bacteria, the choice is made by means of local bacterial selections, thus amplifying the number of favorable genes and shunting them along circuits of genetic exchange to the strains requiring such information. This biological communications network, which possesses more basic information than the brain of any mammal, functions in a manner that sometimes resembles human intelligence (47). For example, man uses tools when he requires them but does not carry all his tools at all times. Bacteria temporarily carry their small replicons containing typically bacterial tools, that is, genes coding for certain enzymes. These shared small replicons can readily be exchanged for others if circumstances favor bacteria with different plasmids or prophages. A second analogy concerns the nature of the transmission of information. Favorable genes appearing in a bacterium, generally by the insertion of a prophage or plasmid, are transmitted directly to the cell's descendants by ordinary asexual mechanisms of heredity, and to neighboring strains by genetic exchange. The relevant information follows a two-fold path. A similar path occurs for technical improvements, developed by humans, that are transmitted not only to children but also to neighbors. This phenomenon has enabled mankind to undergo a continually-accelerating technical evolution.

Manifestations of the information exchange in the bacterial world. In recent years, we have seen unequivocal proof of the manifestation in nature of complex and extensive functions of bacteria, involving the participation of many different strains and recurring on several different occasions. The best-known case is the successive acquisition by bacteria of resistance to various antibiotic drugs, one after another, a phenomenon that independently repeated itself in five or six major geographic areas of the globe (17,57).

Originally, the great majority of pathogenic Gram-positive bacteria and several Gram-negative strains showed no resistance to penicillin. Today we know that the widespread use of penicillin caused the mobilization, principally through transduction in Gram-positive bacteria, of several genes conferring resistance to penicillin. These genes were probably initially present in soil bacteria. After being carried through circuits of exchange by temperate phages, these genes eventually reached staphylococci in hospitals, the bacteria most heavily exposed to penicillin. As was the case for many other antibiotics, a beautiful solution was found involuntarily but infallibly by the bacterial community. The gene that was eventually transmitted to the strains continuously subjected to a given antibiotic carried the genetic information enabling those strains to synthesize an enzyme that specifically digests that antibiotic.

This phenomenon was reported for several other antibiotics

that were discovered and applied at various times. In other cases, the resistance gene stopped the drug's action by different mechanisms. In Gram-negative bacteria, the acquisition of drug resistance was accelerated by conjugation plasmids called R plasmids. A eukaryotic species needing a similar enzyme might have required approximately one million years to synthesize it by means of random mutation attempts. The immense reserve of genes of all types and the infallible mechanisms of selection, amplification, and transfer of such genes in every instance of necessity has enabled bacteria to provide specific means of defense for pathogenic strains in nearly all cases, and generally within a few years of being challenged by the use of a new product. The fact that the agent of syphilis is still sensitive to penicillin may prove that this old parasite never acquired, or lost, the ability to exchange genes with other bacteria and that its capacities are severely diminished because it does not benefit from the general, complex functions of the bacterial world.

Phenomena similar to the acquisition of resistance to antibiotics have also occurred in soil bacteria. The widespread application of numerous pesticides and fertilizers has in many cases produced an antibacterial side effect. Using the same mechanisms, involving their biological communications network, bacteria constantly threatened by these substances have finally received the gene or genes providing the formula for the synthesis of the specific enzyme capable of neutralizing or even digesting the organic substances that were initially toxic to them.

These infallible responses to massive applications of antibacterial drugs in human and veterinary medicine, particularly in animal husbandry, as well as to the use of many agricultural products, show that the biological communications network of the bacterial world is not a figment of the imagination but a reality manifesting itself in ways that profoundly affect our lives.

Favorable consequences for the biosphere

Bacteria, by diversifying and maintaining a state of permanent competition, have finally managed to exploit all ecological niches. Bacteria are present in all environments where eukaryotes are found, but the opposite is not true. There are still niches inhabited by bacteria that eukaryotes cannot occupy, such as hot springs, extreme desert soils, and evaporate flats. All possibilities for prokaryotic life seem to have been realized. Through its countless varieties of cells, the bacterial entity has been able to adapt to all situations that are even slightly favorable to the life of some of its cells. The selective transmission of small replicons probably helped in most of these cases.

This basic activity has many applications, one of which is the most appropriate distribution of the genes for the enzymes required in every corner of the Earth that has facilities for bacterial life. Not only will adequate strains replace less adequate ones, but very well-adapted strains that require a new gene will obtain it and continue to be dominant in their favorite environment. The bacterial planetary entity is permanently able, at its own option to carry out any redistribution of cells, shared small replicons, and isolated genes necessary for the existence of the greatest possible number of bacteria. The result is an amplification of the quantity of all living matter, including eukaryotes, and an acceleration of all biological cycles. A very sophisticated type of symbiosis with eukaryotic life has thus been formed to which global bacterial functions, through their untiring activity, contribute a capacity for correction and adaptation, thus maintaining the biological component of our planet in an optimal state (30,34,35).

Coordination and regulation at the level of the global bacterial entity. Many biologists still tend to regard bacteria as unpredictable elements causing disruptions in the equilibrium of the biosphere. Each single strain does, in fact, tend to carry out unrestrained reproduction under favorable conditions, generating population explosions usually followed by periods of scarcity causing the death of many bacterial cells. Many misinformed individuals fear that our planet will fall prey to a newly-modified type of bacterium whose frenetic multiplication could bring about total destruction. This possibility, however, is not compatible with the fact that in general the over-all activity of bacteria results in harmony and stability. These favorable conditions result from the continuous balanced activities of various strains from any team that utilizes environmental substrates as soon as they become available. This is done with such a high efficiency and yield, due to the extreme specialization of bacteria, that bacteria from another environment normally do not have the slightest chance of usurping those that are already entrenched. In the rare cases where such a strain manages to implant itself, it, in turn, becomes an element of stability that can be supplanted only with very great difficulty. Substrates favorable to bacteria never accumulate in nature because they are utilized as soon as they become available. This efficiency of indigenous bacterial microbiota results from the extreme specialization of various strains and from continuous natural selection. In the autonomous functions of strains and teams, a given group of bacteria that performs its work efficiently and as rapidly as possible is in no danger of being either replaced or modified by genetic exchanges that cannot contribute to its improvement. Selection will thus usually work against bacteria that have received new genes. Highly successful bacterial strains are not threatened by the many changes that commonly occur in the bacterial world. Any upheaval is met by a continuity of the efficient functioning of the bacterial world so as to maintain its very ancient equilibrium.

The characteristic mechanisms of coordination and regulation of the global bacterial entity generally favor the success of the best adapted strains, and this property contributes greatly to the stabilization of the biosphere. These general mechanisms, which operate on a global scale, enable strains that are well adapted to their environment to adapt to major environmental changes by means of a minimum of genetic modifications. Thus, the extinction of very highly-specialized bacterial strains, which constitute the majority of the population, is prevented. The mechanisms ensuring these functions are always in existence. Their elements operate at a very reduced rate when no need for specific exchanges of genes exists. Their potential capacity is stimulated and directed towards logical solutions by significant and relatively prolonged environmental modifications. It is when these conditions affect a large area or even the whole Earth that the mechanisms for genetic exchanges possessed by every bacterial cell, mechanisms that appear to reflect a capacity for what might be called genetic altruism, find full justification.

The constant sending and receiving of information molecules, particularly genes, as a result of mechanisms of genetic exchange, is probably much less frequent in eukaryotes. Even in bacteria the likelihood of useful exchanges for a single cell during several generations is practically nonexistent. However, the intense competition between bacteria in nature promotes the reproduction and diffusion of those bacteria that possess the gene necessary for adapting to new environmental conditions when they arise. The transfer of this gene to other strains is then favored and they, in turn, will favor transfer. This transmission of favorable genes, which are always present somewhere in the bacterial world, to strains requiring them is an economical method of coordinating the functioning of the global entity of bacteria, a method that prevents continual upheavals in the equilibrium of populations leading to the disappearance of highly specialized strains. Instead of eternal new beginnings, there is an eternal recycling of the most successful formulas. Coordination takes place primarily through the transfer of shared small replicons (prophages and plasmids) which diffuse along all possible circuits for gene exchange linking the innumerable bacterial strains in nature. The final result is an optimal and permanent redistribution of bacterial enzymes effected through a series of nearly imperceptible changes, involving very few genes at a time, that closely follows the availability of substrates. Although the apparent waste of useless phages, unused surface receptors, and bacteria destroyed by bacteriolysis may penalize an infinitesimal minority of bacteria, the mechanisms involved ensure the survival of the planetary bacterial entity while maintaining the high degree of specialization of all of its component cells. This phenomenon takes place by virtue of a redistribution of bacterial genes among various strains whenever required. Natural selection drives the process, providing both impetus and direction. The basic mechanism consists of small genetic modifications, before transcription. In animals, coordination and regulation phenomena are carried out principally by post-transcriptional modifications of messenger RNA. Both types of superorganism thus avoid regulation by transcription itself. Such a method would involve too many repressed genes and, in particular, too many regulatory genes. The global bacterial system of coordination and regulation is clearly more economical than the eukaryotic mechanism as it uses very limited means to obtain incredible efficiency. This way of life compensates for and vindicates the fact that each bacterial cell is genetically incomplete and has a rather limited metabolism. By solidarity in associations and gene exchange, these limitations are turned into an asset.

Cooperation with Eukaryotes

Associations between bacteria and eukaryotes exist for nearly all eukaryotes. Many are temporary and limited in their effects at the local level; many more are of reciprocal advantage and may be considered to be ectosymbioses. There are fewer associations of distinct, cultivable bacteria living inside the tissues of the cells of eukaryotes and realizing endosymbiosis. There are many protists, invertebrate animals and leguminous plants, that have symbiotic intracellular bacteria (35). In all these associations, the more adaptable associate is generally the bacterial partner, probably due to the unique capacities arising from genetic exchange in prokaryotes.

Among the most successful ectosymbioses is the one between very complex animals and very efficient teams of bacteria that are inside the digestive tract but outside the tissues and cells. As we saw earlier in this chapter, the ruminants and the bacteria carried in their rumen are a very evident example. Cows, sheep, goats, buffalos and all their wild relatives, and the deer family could not survive in nature without their rumen bacterial teams. The same is true for the bacterial cells involved, as the animal cells of such a ruminant are part of their ecosystem. They are all necessary for such an entity to live and multiply. A similar situation is known in termites, where bacteria provide usable nitrogen and metabolize the breakdown products of wood. There are many more animal-microbe associations where the reciprocal dependency, the solidarity, is not so visible; but when we study many ecological niches we see that most of the living things in them, eukaryotes and prokaryotes, are necessary to realize a local association that is, in the end, balanced and viable.

At the planetary level there is a general reciprocity between plants that realize most of the synthesis of organic matter, through photosynthesis, but need minerals to grow and the animals, fungi, and bacteria that realize the decomposition of organic matter into mineral substances essential for plants. Some strains of bacteria are the only living things able to use nitrogen directly from the atmosphere and thus recycle it for all other organisms on Earth. Many bacteria and other organisms contribute to purify natural water, to stabilize the acidity of our environment, and, probably the most important, to stabilize the present atmosphere to which we are adjusted (30). Bacteria are the most stabilizing and adaptive part in this life-promoting activity.

Finally, a sort of ectosymbiosis is established and maintained with all living organisms in the immediate vicinity of bacteria. The metabolism of bacteria living in close proximity to eukaryotes is, in many cases, complementary with eukaryotic mechanisms, as in the digestive tracts of all mammals. We have already mentioned the decisive role played by bacteria in soil fertility. In their relationships with eukaryotes, bacteria are the more adaptable, flexible, and active of the partners. Their capacity to adapt very closely to environmental conditions guarantees the stability of the biosphere.

Most biologists now accept the idea that very probably eukaryotes appeared by a succession of intracellular symbioses between prokaryotes, starting more than a thousand million years ago. By this time, the bacterial entity had already established an equilibrium based on the very long (1500–2000 million years) survival of its best adapted strains, which themselves then had better chances of persisting. The environment was thus biologically buffered, fostering the appearance of eukaryotes. The development of this initial symbiosis and its many consequences changed the aspect of life on Earth and gave rise to the present living world (32,35,55). The Gaia Hypothesis, which presents the entire biosphere as a life-supporting ensemble, advances the idea that the lower atmosphere of the entire Earth is modified and maintained in a form favorable to life in the same way that a living organism maintains a stable internal environment favorable to its components (30). In supporting our planetary homeostasis, the complex functions of bacteria and their tendency to cooperate with other organisms seem to play the most important role.

CHAPTER 5 *A Very Original Evolution*

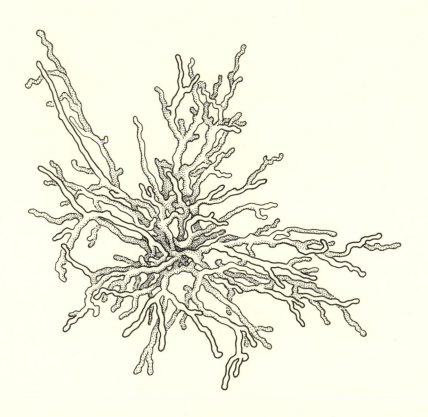

OUR CURRENT knowledge indicates that life had already appeared on Earth approximately three and a half billion years ago. During the first half of that period, the living world very probably consisted exclusively of prokaryote-like cells. It is believed that the appearance of the first living cell on our planet was preceded by many incidences of spontaneous synthesis of complex organic molecules from a high concentration of simpler organic molecules (19,39).

This process of spontaneous synthesis and reorganization from pre-existing organic matter must have unfolded according to the laws of chance. Thus, the phenomenon would have been extremely rare and, in all likelihood, would have occurred only once. The rearrangement of matter into a configuration permitting the localized and correlated synthesis of both protein and genetic matter might have lasted several million years before the appearance of the ancestral cell. Primitive ribosomes had to be surrounded by amino acids, nucleic bases, transfer RNA, a minimum of enzymes, a semipermeable membrane, and RNA bearing the appropriate information for the synthesis of required enzymes. In all likelihood, RNA was the only genetic material in the first cell and there was a mixture of different lengths, mostly short, of RNA replicating molecules. Fragmented RNA viruses might be vestiges of such genomes. RNA presents far fewer possibilities of recombination than DNA, so millions of generations later most of the genetic information may have been transferred in DNA molecules by reverse transcriptase, and RNA templates were used as mRNA.

The First Prokaryotic Cell; the Formation of a Planetary Clone

After another long period of time and, no doubt, after countless failures, the first cell probably divided in some way, producing as offspring two nearly identical cells. Only from this point on was the continuation and propagation of life on our planet assured. It is likely that the parent cell and its immediate descendants had extremely slow metabolisms and that cell divisions occurred only monthly, or even yearly. Despite this low rate of reproductive activity, a favorable environment and the absence of natural enemies made geometric growth possible, and it seems highly probable that within a few millenia the entire Earth was inhabited by this first clone.

> Earth's environment less favorable to other forms of life as a result of intense initial multiplication of bacteria

Eventually, these bacteria depleted the early Earth's vast reserves of prebiotically-produced organic matter, so that there was a rapid disappearance of certain important molecules such as amino acids, other organic acids, and sugars. Initially, only those cells able to survive without some of these substances continued to multiply. Thus, an inexorable change in the Earth's metabolic population structure took place in the next million years. The most highly adapted cells—those capable of continually surmounting new difficulties—must have eventually developed the ability to synthesize enzymes that were either capable of exploiting other nutrients or, perhaps, of synthesizing the important or essential molecules that were being depleted. The bacterial population was rapidly rendering the environment far less favorable for any possible, but improbable, newly-formed cell than it had been at the time of the appearance of the original ancestral cell. This condition would have eliminated the likelihood of a repetition of the slow, frail process of the creation of the first cell and its subsequent survival in isolation, waiting to discover the process of cell division. The first cell's countless descendants would soon have deprived our planet of ready-made nutrients. Consequently, it is most probable that all the cells on Earth are descendants of a single ancestral prokaryotic cell. It can be stated unequivocally that during the first few thousand years after the appearance of life, these constantly replicating cells formed a typical clone.

Permanent selection

Natural selection initially favored those types of cells that multiplied most rapidly. However, the decrease in available organic matter generated a selection based on competition for the best use of the scarcest substances, or for their synthesis. The least adapted cells, with absolute requirements for these substances, disappeared, while the better adapted cells were able to survive.

Unity of the Bacterial Clone; Preservation of the Common Genome

Although changes probably occurred in very small increments, each requiring many generations, the primitive bacterial clone became increasingly diversified metabolically and was probably in danger of splitting into countless unrelated, isolated strains (7,56).

Reunification of the planetary clone through genetic exchange

A reunification of the clone very probably took place through several types of events, each appearing at different times. It seems that all these events led to the same result: the increasingly efficient exchange of genes between bacterial strains, corresponding to a pooling of all genes. It is likely that cell membranes were, from the very beginning, permeable to DNA from surrounding cells of other strains, whether viable or dead. Such small fragments of foreign DNA were probably sometimes inserted by recombination in place of a similar zone on the genome of the receptor bacterium (exclusion–modification enzymes were probably not present in early bacteria). In this way, all those cells that were able to benefit from this phenomenon (called transformation) became the originators of the genetic solidarity that is a characteristic of prokaryotes.

The establishment of a potential genome shared by all bacteria

This pooling of genes resulted in the development of the first common potential genome of all bacteria, and this genome has since continued to grow. The evolution of bacteria into specialized cells by means of continually improving adaptation to any one of the existing metabolic possibilities was thus considerably accelerated (Figure 5–1). Strains which had evolved through their own means were able to receive new genes refined by other strains, and if the new genes proved useful, to utilize them immediately instead of spending millions of years attempting to synthesize them "de novo." This capacity allowed such strains to exploit new ecological niches relatively quickly.

The essential role of shared small replicons or visiting genes

Prophages probably appeared many million years after the phenomenon of transformation. Today, all bacteria seem to possess them. One such shared small replicon can visit a bacterial strain for several generations; it may carry several genes from another strain. These borrowed genes will be expressed; that is, the information they contain is read and used to direct protein synthesis by the host cell that thereby acquires the coded enzymes. This type of genetic exchange, called phage conversion, is much more widespread than transformation. These phages can also carry out transduction, whereby a much greater number of genes can be transferred from one bacterium to another in their natural sequence. The dissemination of phages may have brought about, by transfer, the proliferation of the most successful formula for ribosome synthesis and perhaps the best formula for cell division.

As the number of bacteria on Earth increased, competition among them dominated, and still governs the evolution of their vital processes. Survival became increasingly assured only for those cells with the most active metabolisms and shortest cycles of division.

Appearance of the cell wall

The presence of a peptidoglycan cell wall in several bacteria favored the acceleration of metabolism and cycles of division by

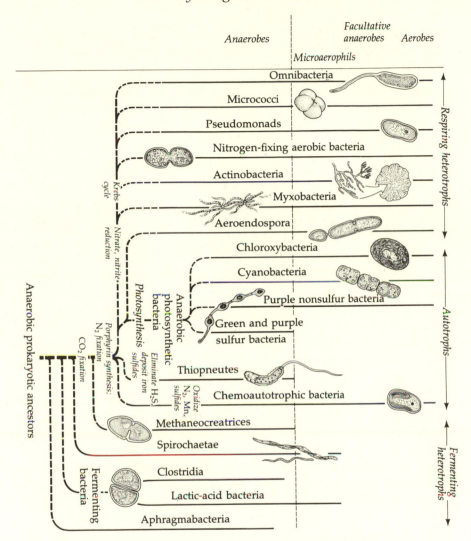

Figure 5–1. A prokaryotic phylogeny. (From *Five Kingdoms: An Illustrated Guide to the Phyla of Life on Earth* by L. Margulis and K. Schwartz. Copyright © 1982 by W. H. Freeman and Company. All rights reserved.)

permitting very high intracellular concentrations of molecules. Without this wall, such a high concentration would have caused cells to explode as a result of great differences between internal and usual external osmotic pressure. The genetic information for synthesizing this type of cell wall very probably spread progressively, most likely through transduction of the relevant genes, to an increasing number of bacterial strains, and finally to most of present bacteria. Only Archaebacteria kept another, although similar, chemical constitution for their cell walls.

*The appearance and abundant repro-
duction of photosynthetic bacteria*

The appearance of a new type of bacteria, capable of photosynthesis, probably occurred between 2.5 and 3.5 billion years ago. These bacteria were able to utilize solar light for the synthesis of organic molecules from carbon dioxide and water. This was one of the most important advances for life on our planet. In contrast with other sources of energy, which were much more limited and unequally available in space and time, solar light is general and permanent. The photosynthetic bacteria needed only the necessary elements in organic or mineral form to synthesize new cell material. Naturally, they could multiply mostly at the surface of expanses of calm water; on the surface of normal soil there is not enough water for them where sunshine is present. So it was very likely that there were few bacteria in the soils of the old continents, probably cyanobacteria, before the advent of eukaryotes. The number of photosynthetic bacteria in general increased until they became the most important metabolic group (36). A further achievement in the same line, of greatest consequence in paleobiology, was the later advent of cyanobacteria (blue or green algae), such as *Microcoleus*. Most of them are entirely autotrophic cells, and can utilize atmospheric nitrogen. A few types of these cells began to liberate oxygen from the carbon dioxide and water that was transformed into organic molecules during photosynthesis. This process led to the first appearance of oxygen in sediments and later in the atmosphere, about two billion years ago, and its concentration increased, perhaps precipitously, until it reached its present level.

As this momentous change gradually took place over the course of millenia, most of the bacteria exposed to air were forced either to adapt to the toxicity of oxygen or to die. The exchange of genes, by then commonplace, probably permitted the diffusion of genetic information favorable to oxygen tolerance among many bacterial strains. A selection for strains with the best systems of genetic exchange was also operating, since strains with optimal systems would have been able to adapt more quickly to changing conditions. Thus, the genetic solidarity of bacteria increased, contributing significantly to the unity of their planetary clone. Eventually, a few bacteria even succeeded in synthesizing the genes corresponding to enzymes that enabled them to utilize free oxygen from the new atmosphere.

This respiration pathway was more efficient than earlier methods of energy metabolism. These valuable genes providing better adaptation to aerobic conditions were acquired by many strains, most probably through one of the methods of genetic exchange discussed earlier.

Life in general was now made easier by the presence of oxygen, which was originally toxic for cells. Oxygen molecules combined to produce an ozone layer in the higher strata of the atmosphere. The ozone blocked an important part of the ultraviolet solar rays that would be lethal to unprotected cells.

Appearance of Eukaryotes; Divergent Evolution for Both Groups

The end of the period of evolution we have been discussing, around one and a half billion years ago, marks the beginning of another great biological adventure: the appearance of eukaryotes. It is likely that the new era began either with an oxygen-tolerant anaerobic bacterial strain or with an aerobic strain having a low energy yield. The strain might have lacked a rigid cell wall and was most likely very competitive, being the only one capable of phagocytosis (envelopment and digestion of cells or other particles). There is no evidence of phagocytosis, however, in modern prokaryotes, so this bacterial lineage seems to have been replaced by its eukaryotic descendants. Through phagocytosis, one cell of this strain may have acquired as endosymbionts a few aerobic bacteria that possessed the most efficient type of energy metabolism, the one that uses free oxygen as the final hydrogen receptor. Alternatively, such a bacterium lacking phagocytosis may have been penetrated by a *Bdellovibrio*-like respirer. The *Bdellovibrio* predators became endosymbionts. These endosymbionts most probably constituted the precursors of mitochondria, now present in most eukaryotic cells (32,35,40). In the ancestral eukaryotic cells, with the important advantage of available energy, the number of genes in the future nucleus could be increased progressively, probably through the retention of genes from visiting prophages and plasmids (48). This process eliminated from these cells their ability to carry out genetic exchanges with surrounding bacterial strains, since the occupation of an attachment site by any small replicon usually blocks the implantation of a similar replicon, and the presence of one type of small replicon in a bacterium usually prevents the multiplication of a similar one. The resulting eukaryotic cells no longer had access to the immense common genetic potential of their bacterial ancestors (Figures 5–2, 5–3, 5–4).

The evolution and differentiation of each eukaryotic cell was limited to the exploitation of its intracellular genome. This genetic isolation accelerated the formation of species and their diversification, each new species retaining no gene exchange bonds with those from which it had separated. A few hundred million years later, the advent of meiotic sexuality in eukaryotes, however, lessened their genetic isolation to some extent. In meiotic sexuality, a new organism is formed through the fusion of two nuclei, one from each parent, each containing a randomly-selected half of the parent's genes. The size of the potential genome was thus extended to the limits of the particular species, where a species is defined precisely as that group whose members are capable of producing fertile descendants through a sexual process. The advent of sexuality radically changed the lifespan of eukaryotic clones by interrupting the genetic continuity at each generation. Whereas the unified, complex bacterial clone of the original type has existed for approximately three billion years, eukaryotic clones correspond to individuals, and, except in rare instances of parthenogenesis, they are essentially mortal (Figure 5–4). Thus, strangely enough, in our eukaryotic world sexuality is linked with death.

As a result of the divergent evolution of eukaryotes following their origin, the protoctists—algae, protozoans, slime molds, and so on (58)—and later animals appeared. Eukaryotic red and green algae were probably first formed only about seven to nine hundred million years ago when eukaryotic protists took photosynthesizing, oxygen-liberating cyanobacteria as endosymbionts. These cyanobacteria were probably of the Prochloron type in the case of the green algae and of plants. There are indications that more than one successful symbiosis involving cyanobacteria

3–3.5 billion
years ago

1.5 billion
years ago

1 billion
years ago

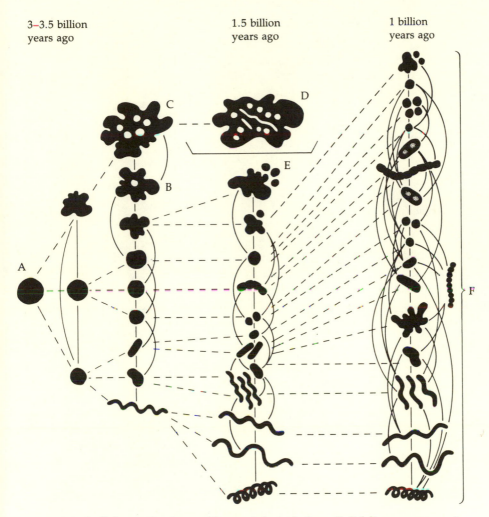

Figure 5–2. Schematic view of bacterial evolution. Solid lines represent
the capacity to exchange genes.

A: First cell on Earth.

B: Prokaryotic cell, possibly slightly phagocytic.

C: Prokaryotic cell, possibly strongly phagocytic.

D: Ancestral eukaryotic cell containing the ancestors of mitochondria as
endosymbionts.

E: Cells of a mycoplasmal type, too large to survive competition with
eukaryotic cells, and thus without descendants.

F: The bacterial world at present, constituting a superorganism con-
solidated by countless circuits for the exchange of genes.

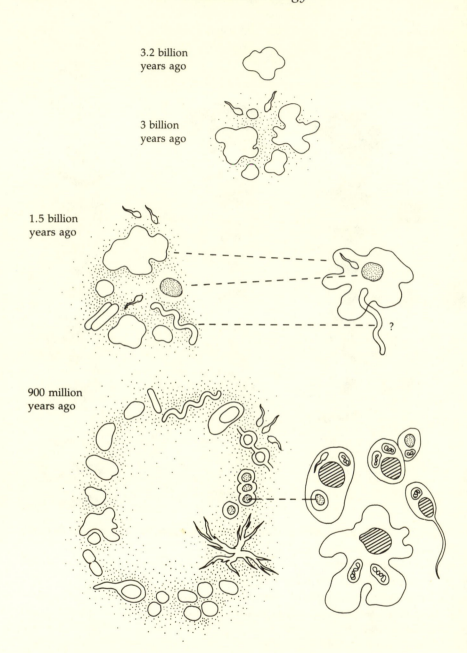

Figure 5–3. The probable origin of eukaryotes. The dotted areas surrounding the cells represent bacterial gene exchanges. Dashed lines in-
dicate the symbioses made by different bacteria that thus abandoned the
bacterial gene pool.

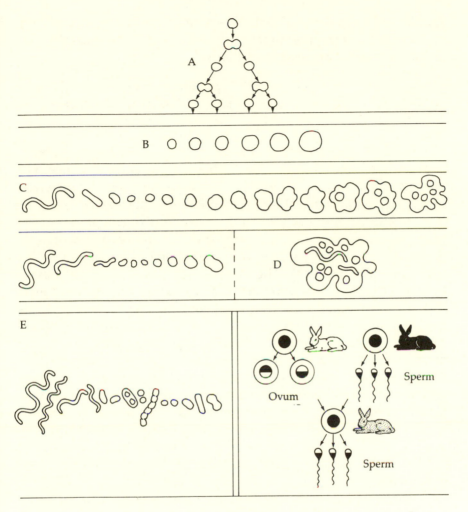

Figure 5–4. Genetic continuity of bacteria compared to genetic discontinuity of eukaryotes.

A: Bacterial division into nearly identical offspring cells (more than 3.5 billion years ago).

B: Structural and metabolic diversification in bacteria, but genetic continuity from one generation to the next.

C: Maximal diversification in bacteria, including consortia of phagocytic bacteria (1.5 billion years ago).

D: The first eukaryotes, formed by successive endosymbioses. Their success eliminates similar bacteria.

E: After the appearance of meiotic sexuality, there is genetic discontinuity between generations in eukaryotes. Ancestors of eukaryotes all died; ancestors of bacteria did not.

probably occurred (33,35,36). The algae have since successfully replaced by competition most of the photosynthetic bacteria, and probably were at the origin of plants.

Equilibrium between different metabolic groups of bacteria disturbed by successful eukaryotes

Prior to the appearance of eukaryotes, all types of metabolism on Earth were carried out by bacteria. A stable balance had been reached in the distribution of strains carrying out different functions. In particular, there was most probably a higher proportion of autotrophic bacteria (not dependent on organic matter) than heterotrophic bacteria (dependent on available organic molecules and active in their decomposition of organic molecules into mineral molecules). The appearance and subsequent development of eukaryotes, and particularly of algae and plants altered this stability, replacing it progressively with another one that has persisted and presently enables the Earth to nourish the greatest number of living cells yet attained. The eukaryotes' success is explained by their ability to exploit areas where the multiplication of prokaryotes was limited by their inherent nature. A growing division of labor between the old and the new group, arising from simple competition, took place quite rapidly. These two major groups of living creatures evolved divergently as their differences and the supplementary nature of their roles became increasingly accentuated. Bacteria developed some of their own features to a greater extent.

Plants and animals present one more genetic discontinuity found in eukaryotes, in addition to that due to the death of each clone. These eukaryotes possess highly-differentiated somatic cells which are usually unable to transmit their genetic material to descendants.

Marked differences between prokaryotes and eukaryotes

Eukaryotes possess remarkable possibilities of intracellular differentiation. The mechanism that accounts for this is the derepression of a portion of the vast intracellular reserve of repressed genes. Derepression allows the expression of the cell's genetic message to unfold right down to the synthesis of an enzyme or other macromolecule. During its lifetime, any mammalian cell expresses only one twentieth of its genes. The other genes are repressed and therefore unused. For example, the genes that allow our eyes to be photosensitive exist in all other cells of the body without manifesting themselves. Throughout the course of evolution, this differentiation mechanism favored the appearance of multicellular eukaryotes, in which a single clone, whether plant or animal, was able to form tissues and organs containing highly specialized and closely cross-supporting cells. For example, cells of specialized tissues that are unable to nourish themselves must be provided for by the whole organism. The same mechanism of selective derepression has enabled the peripheral cells of multicellular eukaryotes to form a continuous specialized layer insulating the organism from the external environment, except where contact is provided by a few effectively protected openings. Skin and its analogues (such as bark and cuticle), delimit and protect the internal environment against evaporation and keep out competing microorganisms, harmful physical agents, and chemical substances (36). This same capacity for the formation of specialized cells has enabled plants to elevate themselves from the soil by means of stems or trunks, and to send their roots deeply into the soil. Thus equipped, plants were able to leave their aquatic environments and colonize the land. The chloroplasts, not-so-ancient descendants of cyanobacteria when the land was first conquered, were supported by the typically eukaryotic assemblage of cells forming branches so that they were perfectly exposed to sunshine. The roots went deeper under the soil's surface to find moisture and minerals that were pumped to help the photosynthesis of the recently co-opted chloroplasts. Nitrogen-fixing bacteria had to accompany the plants' conquest of land, as only bacteria can use atmospheric nitrogen to form organic compounds. The life-carrying capacity of our planet was thus increased many times. Most of the photosynthesis on Earth is now carried out by plants and red and green algae, not by bacteria. Following the algal distribution came animals, endowed with locomotory apparatus that allowed some to live on land, others to remain in fresh or salt water, and still others to exist in both environments. Birds, and many insects as well, were able to rise into the air. Their digestive tracts are always full of billions of bacteria: nutrients are brought in regularly and the humidity and temperature there is favorable for them. For such intestinal bacteria, the host animal is a mobile harvester of food, provided with the facilities of a fermentor. Thus the Earth as a whole became able to accommodate an ever-increasing number of living cells.

Last major stage in the adaptation of the bacterial world

During the time from the first appearance of eukaryotes until the colonization of land by plants, and later by animals, the members of the planetary bacterial clone underwent a final major phase of adaptation. Through active specialization they availed themselves of areas where eukaryotes had no obvious superiority. They developed additional specializations in order to take advantage of the presence of eukaryotic wastes in soil, in beds of slow-moving rivers, in lake and sea bottoms, and in the digestive tracts of animals. The principal role played by bacteria since the flourishing of eukaryotes has been the transformation of dead cells and excreta into minerals assimilable by plants. Heterotrophic bacteria once again outnumbered autotrophs, as they had when life first appeared on Earth.

In order to perform this major reorganization, the planetary bacterial clone had to be able to benefit from the permutations of genes among strains, a process that had occurred in every period of crisis or crucial adaptation. These permutations were directed by countless local selections, using the bacterial communications network described previously. The bacterial world was thus able to carry out this final major adaptation successfully, and was itself able to proliferate as a result. This adaptation also constituted the final episode in the reunification of the global bacterial clone. Again, the strains best adapted for genetic exchange survived more successfully than others. These events seem to have occurred as if the most efficient method of genetic exchange—conjugation—had appeared at that time or shortly before. The fact that conjugation is almost exclusively limited to Gram-negative bacteria (which constitute about half of all bacteria) indicates that it is probably a more recent development than are the other methods of genetic exchange (13,17).

A Corrective System for Evolution

An increased capacity for mutual aid among bacteria by genetic exchange enhanced their ability to correct the evolution of each strain and to carry out, when necessary, a temporarily reversible evolution. Each bacterial strain was able to develop its specialization for a given biological role to an advanced degree. This superspecialization did not result in ultimate extinction, as happened repeatedly in highly-specialized eukaryotes. Both the superspecialization of bacterial cells and the capacity for shorter generations were positive characteristics in the fight for survival. These possibilities were both served, on one hand, by the acquisition by most bacteria of a cell wall (except the mycoplasmas and temporary L forms) and, on the other hand, by a progressive decrease in the strict minimum number of intracellular genes, in particular those of the genophore.

> *Greatly reduced intracellular genome of all bacteria, favoring a unique type of differentiation*

The consistent very small size of the intracellular bacterial genome is one of the most surprising characteristics of bacteria today. Moreover, it has been shown experimentally that any accumulation of prophages or plasmids penalizes a bacterium. Prophages or plasmids tend to be replaced by others, rather than be accumulated (27,47,49). As we have seen, the intracellular prokaryote genome is in fact incomplete and cannot undergo on its own the type of evolution or differentiation found in eukaryotes. The principal type of differentiation in bacteria occurs through the differential distribution of bacterial genes from the planetary potential gene pool among various bacterial strains. This condition arises from the unique evolution of the bacterial clone that acquired more and more genes and specialized strains, all as parts of the full entity. Most redistributions of genes were generated by major changes in environmental factors. Differentiation by means of a redistribution of genes among strains occurs each time it is needed, most often with the help of prophages or of plasmids.

Prophages and plasmids have also evolved, both independently and by acquiring genes produced on other replicons. Each has perfected a mechanism of genetic transfer with maximum economy of energy and optimum frequency, according to the given circumstances. They adapted either to facilitate visiting new strains or to limit their scope of action. Since most of their genes are repressed, prophages or conjugation plasmids, in contrast with bacterial genophores, were able to harbor new genes undergoing modification during the lengthy assembly of successive sequences of appropriate nucleotides. The shared small replicons became constituents of the planetary clone, and essential to its unity in their role as exchangers of genes.

The evolution of bacteria closely followed conditions on Earth. The Earth's physical contours provided a structural support for bacteria, similar to the way a trunk supports a vine. Water in oceans, seas, lakes, rivers, and rain constituted a basic substance absolutely essential to every bacterium. The circulation of air and displacement of water ensured the renewal of primary elements such as carbon dioxide, oxygen, and nitrogen. Bacteria adapted to this reality and even modified it to improve its life-supporting possibilities (30). The appearance and development of plants and animals, themselves a by-product of bacterial evolution, increased the number of available ecological niches. Through a redistribution of cells following countless subclonal selections and, at the end of this redistribution, a distribution of genes among strains, the bacterial clone could, whenever the need arose, provide to each biological sector an optimal combination of enzymes for its current conditions. The local bacterial teams finally reached an exceptional level of cooperation among their different cells, and realized extremely complex metabolisms, much as organs do in an animal. Thus, the equivalents of the components of one organ (but dispersed, as are our blood cells) are assembled in any viable corner of our planet, and their composition is precisely adjusted for local conditions. These arrangements are in a dynamic state, ready to change or to start in a new way if conditions around them change. Some of these changes are repetitive. The endless succession of cold, temperate, and hot seasons in nature produces a parallel set of changes in bacterial life. In desert regions, for example, the lack of water inhibits the reproduction of bacteria for long periods of time. In other regions, rainy seasons alternate with drought. In all cases, local bacterial activity resumes as soon as environmental conditions become favorable once again. The capacity of bacteria to form spores particularly favors these vital rhythms.

> *Coming of age of the bacterial world: an organized, diversified, genetically unified entity able to improve the environment and direct its own evolution*

Most bacteriologists seem to agree that after such a long evolution, which also constituted an ontological development (akin to animal embryogenesis) taking place inside a genetically unified superorganism and lasting several billion years, the bacterial entity seems to have synthesized all the possible genes necessary to its various cells. Thus, the venerable global bacterial clone, composed of different specialized strains and covering nearly all possible variations, can be said to have attained maturity. We may consider its ontogeny (development of a single organism) and phylogeny (evolutionary development) to have been the same phenomenon. Its harmonious development, including the autonomous or correlated evolution of its components, complementing that of the eukaryotes in a mutually-supportive loose association, seems to support the Gaia Hypothesis (discussed in Chapter 4) in which all living beings are presented as a stabilizing ensemble for the entire inhabitable part of our planet, including its atmosphere (30). The prokaryote part of this ensemble can certainly be considered to form the most closely-knit entity of the biosphere. One general characteristic of bacterial evolution might be that it can be called a "guided evolution." It is a fact that in every local or widespread selection of the fittest cells, bacteria with the better capacity for gene exchange generally perform better and therefore survive. For instance, we have seen recently in the acquisitions of drug resistance by pathogenic bacteria that a few types, like the *Pseudomonas*, became winners. In many situations, the most successful strains were equipped with a reserve of small replicons that was larger than usual. Thus, there seems to be a permanent selective pressure favoring strains that will be the most "sociable," the most genetically communicative. These cases reflect an anti-extinction activity that rescues specialized strains confronted with threatening situations. As we have seen, this active intervention in bacterial evolution played an important role in its long development. Thus, the more flexible strains that were capable of exchanging genes were favored by this global activity that we have compared to the communications network of a vast biological computer. The more isolated, independent strains could not be helped during transition periods and probably disappeared. In this way, general bacterial activities seem to have directed the evolution of all bacteria towards forms with greater capacities for gene exchange, and thus towards a more unified bacterial world.

A Non-Darwinian Form of Evolution

Bacterial evolution differs from that which Darwin described for the eukaryotes, the only creatures with which he could be familiar in his time. For biologists, eukaryotic evolution seemed to be unavoidably related to the genetic isolation of different species, its general movement being always away from their origin and generally away from other species. In contrast, bacteria lacked genetic isolation and consequent differentiation into species. Their evolution evolved inside their own common particular clone, with a huge available gene pool. Different lines of specialized metabolism had their own evolution in the total bacterial community. Groups of metabolic pathways, maintained by selection, probably had more continuity than any given cell lineage (36). The genetic solidarity of bacteria gave them the capacity to correct, temporarily or permanently, characteristics that became unfavorable following a change in the environment. Competition thus contributed to the stability of the most highly-specialized strains, maintaining their advanced degree of adaptation to their respective environments. With a common gene pool, all bacteria evolved as an original type of clonal, dispersed superorganism whose constituent cells displayed a very rich metabolic diversity. Since all bacteria derive most probably from the same parent cell, and from the same increasing gene pool to which they contribute their own genes, all bacterial cells are a part of this original entity. Hence, it may be said that the entire bacterial world has undergone evolution as one individual or as a communicating society, just as mankind undergoes cultural and technical evolution. Bacterial evolution may even demonstrate a deceptively Lamarckian touch: after all, when a strain acquires a new favorable gene and therefore a trait by selection (choice) following the exchange of many genes, this learned information (from another strain) is subsequently inherited by descendants. However, the duration of such a gene will only correspond to the duration of the circumstances favoring it.

CHAPTER 6

Bacteria and Mankind

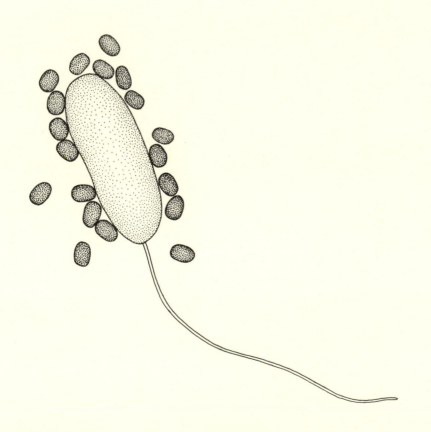

As WE HAVE seen, the appearance, survival, and flourishing of eukaryotes were closely linked to the presence and activity of bacteria. First, and foremost, it is highly probable that the ancestors of eukaryotes were bacteria who eventually formed a symbiotic association, and this association gave rise to the first nucleated cells from which we have descended. It was the life-promoting general activity of all bacteria that later supported the extreme variety of eukaryotes. The relationship of bacteria to stone-age man was essentially the same as it was to other mammals. Man benefited from the general activities of bacteria and, from time to time, suffered the unfavorable consequences of the activities of certain pathogenic bacteria and of those responsible for the degradation of foods he was trying to preserve. It was probably during the Paleolithic period, lasting about seven hundred thousand years, that many bacterial constituents of the normal flora of skin and the digestive tract, common to several primates, were progressively modified in man and eventually gave rise to several bacterial types that are pathogenic exclusively to humans. Other bacteria remained pathogenic to a wide variety of primates, including humans. It goes without saying that our early Paleolithic ancestors knew nothing about the bacterial world. At most, they may have discovered empirically certain methods of food conservation, such as the drying of meat.

From the Beginning of Agriculture Until the Discovery of Bacteria

From the beginning of the Neolithic period, about 10,000 years ago, the population of the human species increased greatly. Potentially fertile lands were gradually transformed into cultivated fields and pastures. The domestication of several animal species progressively replaced wild animals as a dietary source of protein. Agglomerations of domestic animals and humans, living first in villages and later in cities, created pollution problems such as the accumulation of wastes and contamination of water. As we have already seen, soil and water bacteria have always contributed to solutions of these problems. For the last three thousand years, a large proportion of the fertile lands on our planet have been cultivated. Bacteria have had no difficulty in adapting to these changes. They continue to ensure the fertility of soils, the decomposition of dead organic matter, and the natural purification of wastes and water. The global population of the intestinal bacteria of domestic animals and man paralleled the rapid population increase of their hosts. It is likely that the development of husbandry favored the exchange of bacteria of normal microbiota, as well as certain pathogenic bacteria, and even more of their shared replicons between domestic animals and humans. Man became familiar with the manifestations of certain bacterial activities. Infections, and highly contagious diseases in particular, had for a long time been treated by empirical means, usually by the isolation of sick individuals. Food conservation techniques were developed and refined empirically, although it was not known that what these procedures did was to prevent the reproduction of bacteria. Cultural customs and religious rites favored certain populations by protecting them against contamination by pathogenic bacteria. Some examples include the preparation of tea, wine, and beer; the cooking of foods; the washing of hands before meals, ritual baths, and other cleanliness practices; waste disposal; and the salting and drying of meat.

Consequences of Industrialization and Scientific Progress

Industrialization greatly increased the productivity of human societies. Medical advances, particularly in the battle against infectious diseases, together with increased capacities of production, generated an unprecedented population explosion. During the last two centuries, the global human population has increased tenfold. Cities contain a hundred times more people than they did a century ago. Despite these high concentrations of populations, which in the past were favorable to epidemics, modern developments in hygiene and, to a lesser extent, medicine have led to the nearly complete elimination of many contagious bacterial infections.

The raising of domestic animals, particularly those intended for human consumption, has become a large-scale industry that involves massive agglomerations of animals. The recent practice of adding wide-spectrum antibiotics to animal feed led within only a few months to the appearance of drug resistance, particularly in intestinal bacteria. However, bacteria that are pathogenic to mammals have always represented an infinitesimal minority of the bacterial world. These cells separated themselves to a large degree from the extensive circuits of information exchange available to the great mass of other bacteria. Strictly pathogenic strains that are exclusively adapted to intimate association with animal tissues seem to have lost all or much of their ability to communicate with the bacterial entity as a whole. Consequently, in most cases they have not acquired drug resistance. Some pathogenic bacteria of the skin and intestines, more exposed to soil bacteria and their genes, have, however, acquired extensive drug resistance as a result of the capacities of the genetic communications network of the bacterial entity. This resistance can be countered by increasing the variety of antibiotics and by a sudden change of the agents used, according to a well-established procedure. In fact, a bacterial cell cannot possess more than five genes conferring resistance to, respectively, five antibiotics without endangering its survival. Moreover, the biological data system of all bacteria requires a period of time ranging from several months to several years in order to counteract the use of a new drug in a given region. We humans, in some cases, can act more quickly!

In the past few decades, man has begun to domesticate bacteria. Progress has been made in the use of bacteria for the industrial production of certain biological substances. Because of the infinite possibilities for the improvement and even the creation by the appropriate gene exchange, of new strains with a specific biological mission, the future of industrial microbiology seems very promising. In addition, biological research makes extensive use of bacteria as experimental models. Another goal involves eventually learning to correct and, much later, to improve human genetics by means of genetic manipulations that might be carried out principally with the help of bacteria.

Human Activities in which Bacteriology Plays an Important Role

Human and veterinary medicine and plant pathology. From a historical perspective, we have seen that the fields of medicine, both human and veterinary, and plant pathology first experienced major progress of global significance as a result of the discoveries of Pasteur. Moreover, of all bacteriological developments, progress in the battle against pathogenic bacteria has had the most substantial immediate effect on mankind, particularly the prolongation of the average human lifespan by approximately thirty years. As a consequence, the Earth's population has increased dramatically. Large-scale epidemics as well as common contagious diseases of bacterial origin have been practically eliminated. This means that young adults in their period of maximum economic productivity face a much lower mortality rate than in the past.

Numerous preventive measures, such as immunization, health control of travelers, sanitary processing of water and foods, as well as antibiotic treatment of clinical infections, have created an effective barrier against serious bacterial infectious diseases. This barrier must, nevertheless, be tended with vigilance. Through the elimination of major bacterial pathogens, man has induced the appearance of new infections by opportunist bacteria that have produced new types of disease that are difficult to fight.

Progress in veterinary medicine against infectious diseases of bacterial origin paralleled developments in human medicine. Many pathogenic bacteria attack both humans and animals, sometimes provoking very different clinical symptoms in different species. Other bacteria attack certain species exclusively. The industrialization of agriculture has also brought about the appearance of opportunist bacteria, causing new types of infections in domestic animals. The pathogenic bacteria responsible for the infections of reptiles, amphibians, and fish are less well-known than those that infect birds and mammals. Our knowledge of the bacteria that infect invertebrates is very scanty, and much work remains to be done in this area of bacteriology.

Although plants are not known to possess an immunological system of defense, they are, with a few rare exceptions, able to resist infection by bacteria fairly well. Recently, the very important role of several pathogenic mycoplasmas in plants has been demonstrated. In general, however, plants appear to be more susceptible to viral and mycotic (fungal) infections than they are to bacterial infections.

Soil bacteria in agriculture, purification of used water, and ecology. We have seen that soil bacteria play an important role in agriculture, water purification, and in the global ecology. Under natural conditions they degrade organic wastes and fix atmospheric nitrogen, thus permitting plant growth. Agronomists and farmers have attempted to exploit these activities in order to increase the productivity of soils. Scientists hope to be able to effect the transmission of genes responsible for the fixation of atmospheric nitrogen to common soil bacteria and possibly even directly to plants. Several bacterial strains contributing to the purification of wastes, sewage, used water, and to the maintenance of an equilibrium favorable to life in waterways, lakes and seas originate from the bacterial teams of the soil. Ecologists and hygienists use specific methods to determine whether the biological equilibrium has been disturbed, in which case corrective measures must be taken.

A number of techniques for the conservation of agricultural products have been developed or improved as a result of a better understanding of microorganisms, and of bacteria in particular. The necessity of feeding an exponentially growing world population has generated extensive research on the possibilities of using microbial agents to transform various wastes into food substances. For example, the transformation into food products of certain petroleum refinery by-products, which otherwise cannot be utilized, has been shown to be technically and economically feasible. In addition, the energy crisis has revived the interest of scientists and environmental protectionists in harnessing methanogenic bacteria for the production of fuel from urban and agricultural wastes. Several experiments have demonstrated that this type of transformation is both technically and economically feasible, although not on a very large scale.

Industry. Many sectors of industry involved in the use and transformation of organic matter employ bacteriological techniques. Sterilization by heating or irradiation eliminates all traces of life in sealed containers, thus preventing harmful bacterial activity. In the food industry, certain unwanted bacterial activities in liquid food products are prevented by pasteurization, which involves heating for a short period of time at a temperature slightly below the boiling point of water. For the conservation of other foods, the inhibition of unwanted bacterial activity is accomplished by refrigeration or freezing, drying, or salting.

The pharmaceutical industry carries out the preparation of several antibiotics and other biological products with the help of specific bacterial strains. It is also hoped that various substances active in the human body, such as insulin, interferon, and several peptide hormones, will eventually be synthesized by bacteria through the use of techniques of genetic engineering to effect the transfer of the relevant human gene to common bacteria. At present, certain enzymes used in cheese making can be obtained more economically with the help of bacteria than, as traditionally, from domestic animals.

The exploitation of large masses of bacteria in the near future for food production and even for a source of energy is a dream that has already found some realization. Our knowledge of the infinite possibilities of transferring genes from one bacterial strain to another points to a very promising future for the domestication of modified bacterial strains for certain industrial uses.

Research. Bacteria contribute in a crucial manner to basic research in biology as well as to applied research in medicine, agriculture, and industry. Because they are autonomous cells that can easily be grown in pure cultures, certain bacteria have served as the ideal laboratory organism in the search for solutions to general cellular or biochemical problems. *Escherichia coli, Salmonella typhimurium,* and *Bacillus subtilis* and their shared small replicons (prophages, virulent phages, and plasmids) have played this role for most research in molecular biology. An accumulated investment of effort has resulted in a better understanding of these particular bacterial strains, and it has become profitable to build on this acquired data and to seek more information on these same strains.

Although research on bacterial teams has been conducted in several areas of human activity, few studies have dealt with the general mechanisms involved in the functioning of bacterial teams, and even fewer have been concerned with the general properties of the planetary bacterial entity. Other methods for the exchange of information molecules among bacteria have probably yet to be discovered. Further studies must be made of the conditions under which all these exchanges are carried out. Most important, the principal methods of communication between the major groups of bacteria—such as the mycoplasmas, archaebacteria, photosynthetic bacteria, and *Actinobacteria*—and the more common bacteria have yet to be established. One possibly fruitful line of questioning might be, does the biological communications network of all bacteria have important substations interconnected by means of intermediate strains?

It is likely that a method will be found, perhaps within the next thirty years, to begin to correct simple human genetic defects by transferring corrective genes from healthy individuals to several cells of afflicted individuals by exploiting as stepping stones the plasticity of all bacterial genetic material. In the much more distant future, it may eventually be possible to make human evolution reversible and even to direct it favorably, as a result of progress in bacterial genetics and in the corresponding genetic technology.

Human Understanding of Bacteria

The new concept of a unitary society of bacteria and its implications, having no equivalent in other living organisms, is presently neither very clearly understood nor acceptable to many biologists. The 1974 edition of *Bergey's Manual of Determinative Bacteriology* states (page 9): "No doubt time will settle the problems involved in making a reasonable arrangement within the Procaryotae. Haste is unwise; all previous classifications seem to have suffered infinite rearrangement due to insufficient information. . . . The new insights are likely to come from a clear understanding, on a comparative level, of the components of the genome of procaryotic cells." (11) The present book makes its contribution by following exactly this line of thought. Most biologists accept the fact that the living world is divided into two basic groups (super-kingdoms): bacteria (prokaryotes) and eukaryotes (14,50,52). Biologists do not, however, assimilate the implications of this discontinuity among living things. Even most bacteriologists are not familiar with the essential role of small replicons, each one shared between thousands of bacterial strains, or, in particular, with the generalized reality of prophages as the most important gene exchangers. A bacterium without shared small replicons is an exception in nature and is probably so temporarily, before the arrival of a prophage or a plasmid. As a necessary complement to all cell activity, there are decisive simultaneous and complementary activities of the prophages, plasmids, and transposons. Not only are they able to help bacteria in time of need but they also improve the capacity for the solidarity of the whole bacterial world, the capacity for solving new problems, and the capacity for evolution, including partially reversible evolution if necessary.

We have seen that bacteria possess a clonal organization, with many different strains, and we know that they are kept united by a shared total genome that has genes available for exchange between strains at all times, not only at reproduction. Since each bacterial cell has an intracellular number of genes hundreds, or even thousands, of times smaller than the number of genes in eukaryotic cells, most of their differentiation, and evolution at the cellular level, requires the help of genes from other strains. The fact that all or most bacteria, different as they are, have the same potential genome or the same gene pool is in opposition to the familiar and entrenched concept of eukaryotic biological species, which is based on genetic isolation. However, most strains of bacteria are rather stable as a result of the action of the long-standing permanent pressures that support success; they thus represent evolutionary peaks. Their success gave them stability during a longer period of time. They reach a stable differentiation inside the clonal planetary superorganism that may be changed easily if new selective pressures appear, as was often the case with acquiring drug resistance. As we have seen, the metabolism of different strains gradually became more and more specialized during their evolution. These specialized strains can be compared to the highly specialized (differentiated) cells of our body; they all have access to the same genes from all bacteria but are using only some of them. The specialized bacteria are not separate biological species, just as the different cells in our blood are not. Bacterial strains are the specialized cells belonging to one superorganism—the planetary bacterial entity—just as our blood cells belong to one superorganism, the body. There are always a few intermediate bacterial strains in nature; they are not as frequent as the more successful types that have been erroneously called species. Some intermediate strains seem to be more frequent in infections since the advent of antibacterial drugs, as medically important bacteria face many powerful selective pressures that reshuffle their genes.

Because many general improvements in bacteria were shared by gene exchange during their very long evolution, we cannot reconstruct their phylogeny. Only the study of the evolution of certain specific metabolic pathways is promising, not that of a cell's specific line (50). The comparative study of homologies in ribosomic RNA of different bacteria shows, however, some encouraging results for an evolutionary approach. It confirms, for instance, that archaebacteria separated earlier from most other bacteria. This group presents some similarities with eukaryotes, which supports the hypothesis that bacterial ancestors of the eukaryotes, most likely those with a large cell, were probably somehow related to the ancestors of archaebacteria. The comparative study of amino acid sequences in proteins, of nucleotides in nucleic acids, and *in vitro* DNA hybridization and comparative studies of the percentages of guanine and cytosine in DNA can offer only partial hints on some probable recent common origins of different bacteria. Such studies confirm the stability of the previous so-called species that we consider to be successful, and therefore stable, differentiated types of cells. The results in some groups of bacteria, such as the Enterobacteriaceae, are more encouraging than in others. Under these conditions, the classification and even the best nomenclature for bacteria at present can only be utilitarian (47). The editors of *Bergey's Manual* state (page 1) that it "is meant to assist in the identification of bacteria. No attempt has been made to provide a complete hierarchy, as in previous editions, because a complete and meaningful hierarchy is impossible" (11). Because the so-called species and genera are rather stable and because the nomenclature and identification methods that resulted are very convenient, based as they are on metabolism and recently on nucleic acid homology, the nomenclature and subdivisions should be preserved. Thus we have an accepted common bacterial terminology that is based on the concept, no longer biologically justifiable, of species in bacteria. We should understand, however, where the practical aspect is important and where theoretical biology has to be taken into account.

All these considerations support the usefulness of a numeric taxonomy for bacteria. This method shows how all known physiological and biochemical characteristics are distributed among the bacteria in a large, metabolic group. It is totally descriptive; it attributes equal importance to any characteristic of a strain and compares them. It is a logical method of identification since it is based on the known genes present in a strain. As we have seen, most of our contemporary bacterial strains are very stable, so numeric taxonomy is on reasonably sound ground.

Biologists increasingly accept the idea that bacterial evolution very probably has lasted more than 3.5 billion years, in natural waters or natural slimes and mud that kept them exposed to high competition and also favored temporarily adapted associations. We have seen that as a successful answer to this challenge, bacteria selected small cells with very limited, essential genetic material, rather incomplete, compensated by a few efficient systems of exchanging genes at any time, mostly through accessory nomadic groups capable of active transfer (prophages and plasmids). Such incomplete cells, each limited to a very narrow metabolism, survived more easily when associating with complementary metabolic strains. Thus, localized teams of different complementary bacteria formed any time that a new niche was open or when an old one changed its physico–chemical conditions. At the local and at more extensive levels, bacterial associations increased the stability, the duration, and the survival of their cells. At the planetary level, we have seen how bacteria modified the Earth's surface, benefiting from its facilities and stabilizing its water and air conditions. A capacity for complex problem solving is manifested by large numbers of bacteria, continent-wide or planet-wide. As a result of their global data communications network, all bacteria behave as members of a single superorganism with specialized cells, and have a general stabilizing and life-promoting activity.

In this rule-respecting, stabilized system, the first eukaryotes were at odds with what had been normal for 1.5 to 2 billion years. The evolution of this illegitimate new branch of bacteria, as they were at the beginning, took an inside-out turn, compared to that of their bacterial ancestors. We have seen how eukaryotes seem to have kept captive all their ancestor's visiting nomadic genes, and thus cut themselves off from the bacterial gene pool (48). They evolved towards genetic isolation and more and more dissimilar descendants, with few facilities for associations among different species. Bacteria, by contrast, continued to keep their frequent gene exchanges in nature. They continued to be association-prone and also made most of the adjustments needed to benefit from the new world of eukaryotes, especially after these upstarts, helped by essential bacteria, conquered the dry continents. The digestive cavities of animals were colonized with appropriate bacterial teams, with mutual benefit in most cases. The evolution of ruminants, for example, met the bacterial adaptation half way with fermentor-like modifications of their digestive tracts.

Most animals accelerate the exchanges of bacterial strains and genes by their mobility. Migrating birds, insects, and fish accelerate the planet-wide, computer-like communication functions of the bacterial world. We may even think of the chloroplasts in plants as successful cyanobacteria that have found a way of being supported in grass blades or tree leaves to benefit from full sunshine and of being offered water and minerals by the roots of the associated plant, all in exchange for a part of the products of their photosynthesis. Naturally, the most important association is between heterotrophic bacteria, which degrade plant and animal dead cells and excreta, and the nitrogen-fixing bacteria on the one hand, and algae and plants on the other. In this association bacteria are fed and can maintain their exceptionally high numbers. Their large numbers in turn allow them to use their communications network to find solutions for the Earth's life-promoting possibilities and to stabilize its environment. In this planet-wide association of all living entities, bacteria provide most of the facilities for association, cooperation, and stability, and we should no longer underestimate their role.

Glossary

aerobic Requiring gaseous oxygen

anaerobic Requiring the absence of gaseous oxygen

antibiotic A chemical agent produced by one organism (typically a bacterium or fungus) that is harmful to other organisms

archaebacteria Methanogens, *Thermoplasma*, salt-tolerant, and other bacteria that share certain sequence homologies in their RNA (5S and 16S ribosomal RNA)

autotroph Organism that uses inorganic compounds to synthesize organic compounds

bacterium Prokaryotic microorganism

cell wall External, generally rigid, structure produced by cells; contains peptidoglycans in bacteria

chlorophyll Green pigment responsible for absorption of visible light in photosynthetic organisms

chloroplast Green organelle containing chlorophyll in eukaryotes

chromosome Intranuclear organelle composed of DNA and protein in eukaryotes

clone Population of cells all descended from a single cell

colony Cells living together in permanent but loose association

conjugation Transfer of genetic material from one bacterium to another by means of close contact or a pilus

cyanobacteria A diverse group of oxygen-producing photosynthetic pro-karyotes

cytochrome Small protein containing iron heme; acts as electron carrier in respiration and photosynthesis

cytoplasm Fluid, ribosome-filled contents of a cell (excluding the nucleus)

DNA Deoxyribonucleic acid; long molecule of paired nucleotides in a linear order; the genetic material of the cell

enzyme Protein functioning as a catalyst to accelerate specific reactions or groups of reactions

eukaryote Cell or organism composed of cells having a membrane-bounded nucleus

exoenzyme Enzyme that is secreted into and acts in the external medium

fermentation Oxidation–reduction reaction in which the terminal electron acceptor is an inorganic compound but not oxygen

genome Complete set of genes present in an organism

genophore DNA molecule carrying the genes of a bacterium or virus

germinate Start to grow by cell division

Gram stain A staining procedure for identification of bacteria developed by the Danish physician H.C.J. Gram (1855–1938). Any given strain of bacteria may be Gram positive, Gram negative, or Gram variable.

heterotroph Organism that obtains carbon from organic compounds

histone Positively-charged chromosomal protein that binds to DNA

humus Organic part of soils, containing the most decay-resistant components of plants such as lignin

immunity Specific resistance to disease in vertebrate animals, resulting from antibodies in sera and newly alerted defense cells

insertion sequences Sequences of DNA base pairs, able to insert in different places in any replicon, common between replicons such that two replicons can integrate to form one

kerogen Nonextractable complex organic polymeric material in soil, derived from humic acids, that becomes a stable component in sedimentary rocks

ligneous substance Material, like wood, rich in lignin, a complex polyphenolic compound

lipid Organic compound soluble in organic and not aqueous solvents

lysis Loss of cell contents by means of rupture of the cell membrane

mesosome Membranous structure associated with segregation of DNA in dividing bacteria

messenger RNA (mRNA) Molecule of RNA containing a base sequence complementary to DNA; carries information for synthesis of one or more proteins

metabolism All the enzyme-catalyzed reactions occurring in cells

methanogenesis Processes producing methane gas, CH_4; usually involves teams of bacteria that include methanogens

mitochondrion Organelle in eukaryotes responsible for processes of respiration and electron-transfer phosphorylation

mitosis Cell division in eukaryotes, producing two identical offspring cells

meiosis Cell division in eukaryotes in which a diploid cell produces haploid offspring cells

NAD Nicotinamide adenine dinucleotide, and **NADP** (NAD phosphate); pyridine nucleotides, energy carriers in cells

nucleotide Single unit of nucleic acid (DNA or RNA), composed of an organic nitrogenous base, deoxyribose or ribose sugar, and phosphate

nucleus Membrane-bounded organelle in eukaryotes, containing DNA

numerical taxonomy A classification of living things based on the number of all common known characteristics of different beings

organelle Organized structure in cells

peptide Compound containing two or more amino acids linked by the carboxyl group of one and the amino group of another

permeases Enzymes of different kinds that catalyze the transport of small molecules

phage (also called bacteriophage) A virus that infects bacteria, sometimes causing their disintegration; temperate phages are the transferable form of prophages

phagocytosis Ingestion of solid particles in which a cell membrane surrounds a particle and engulfs it

pinocytosis Ingestion of liquid droplets in which a cell membrane surrounds a droplet and engulfs it

plasmid Small replicon, usually cytoplasmic; some may integrate into the large replicon

plastid Photosynthetic cytoplasmic organelle in eukaryotes (algae and plants)

prokaryote Cell or organism composed of cells lacking a membrane-bounded nucleus

protoplast Artificially produced wall-less bacterium from a Gram-positive cell

protein Long-chain polymer of amino acids; some are enzymes, others are structural

replicon   Self-replicating DNA molecule. Large replicon = genophore; the constant portion of the genome. Small replicons = other self-replicating DNA molecules, such as prophages, plasmids, the genophore of a virus; the variable portion of the genome.

respiration   Oxidation of organic compounds in which molecular oxygen, or sometimes nitrate or sulfate ion in some bacteria, is the electron acceptor

ribosome   Cytoplasmic particle composed of RNA and protein; site of protein synthesis

RNA   Ribonucleic acid; polymer of nucleotides serving as genetic storage material in some viruses; used in copying and translating genetic information in other organisms

serum   Liquid portion of blood often containing antibodies that confer immunity against bacterial infection

spore   Asexual reproductive cell, capable of surviving unfavorable conditions

strain   Bacteria isolated from nature and purified so that they are all of a single kind

symbiosis   Intimate and protracted association between two or more organisms of different species, sometimes to their mutual benefit

transcription   Synthesis of RNA in which one strand of DNA serves as a template

transduction   Transfer of genes from one bacterium to another by means of a phage that carries the genes

transfection   Acquisition of small replicons as soluble molecules from another cell

transfer RNA (tRNA)   A type of RNA involved in the translation process

transformation   Transfer of genetic information in bacteria by means of free DNA fragments in the medium surrounding the cell

translation   The production of protein under the direction of mRNA that has complementary sequences as DNA

transposon   DNA fragment, containing one or more genes and flanked by insertion sequences, that moves around replicons and to new replicons

virus   Self-replicating genetic element; DNA or RNA, protected by a protein coating, that can be either intracellular or extracellular; infectious in intracellular state

Bibliography

1 Anderson, E.S. (1966): "Possible importance of transfer factors in bacterial evolution," *Nature* 209: 637–638.

2 Anderson, N.G. (1970): "Evolutionary significance of virus infection," *Nature* 227: 1346–1347.

3 Avery, O.T., C.M. MacLeod, and M. McCarty (1944): "Studies of the chemical nature of the substance inducing transformation of Pneumococcal type. Induction of transformation by a desoxyribonucleic acid fraction isolated from *Pneumococcus* Type III," *J. Exp. Med.* 79: 137–158.

4 Backmann, J., O.B. Goin, and C.J. Goin (1972): "Nuclear DNA amounts in vertebrates," in H.H. Smith, ed., *Evolution of Genetic Systems.*

5 Balch, W.E., G.E. Fox, L.J. Magrum, C.R. Woese, and R.S. Wolfe (1979): "Methanogens: reevaluation of a unique biological group," *Microbiol. Rev.* 43: 260–296.

6 Bak, L.A., C. Christiansen, and A. Stenderup (1970): "Bacterial genome sizes determined by DNA renaturation studies," *J. Gen. Microbiol.* 64: 377–380.

7 Barghoorn, E.S. and J.W. Shope (1966): "Microorganisms three billion years old from the Precambrian of South Africa," *Science* 152: 758–763.

8 Barksdale, L. (1959): "Symposium on the biology of cells modified by viruses or antigens," *Bacteriol. Rev.* 23: 202–228.

9 Blouin, G., L. Laplante, S. Sonea, and J. De Repentigny (1965): "Elimination permanente à l'aide d'une acridine, d'indicateurs de la virulence et de la résistance à l'érythromycine chez des staphylocoques," *Rev. Can. Biol.* 24: 223–224.

131

10 Bodmer, F. (1970): "The evolutionary significance of recombination of pro-karyotes," in H.P. Charles and B.C.J.G. Knight, ed., *Organization and Control in Prokaryotic and Eukaryotic Cells,* 21st Symposium of the Society of General Microbiology, Imperial College London, 1970. London: Cambridge University Press.

11 Buchanan, R.E. and N.E. Gibbons, ed. (1974): *Bergey's Manual of Determinative Bacteriology.* Baltimore: Williams and Wilkins.

12 Bukhari, A.I., J. Shapiro, and S. Adhya, ed. (1977): *DNA Insertion Elements, Plasmids and Episomes.* Cold Spring Harbor, N.Y.: Cold Spring Harbor Laboratory.

13 Campbell, A. (1981): "Evolutionary significance of accessory DNA elements in bacteria," *Ann. Rev. Microbiol.* 35: 55–83.

14 Chatton, E. (1937): *Titres et Travaux Scientifiques.* Sète, Sottano.

15 Cohen, S.N. and A.C.Y. Chang (1975): "Transformation of *Escherichia coli* by plasmid chimeras constructed *in vitro:* A review," in D. Schlessinger, ed., *Microbiology—1974.* Washington, D.C.: Amer. Soc. for Microbiol., pp. 66–75.

16 Cowan, S.T. (1970): "Heretical taxonomy for bacteriologists," *J. Gen. Microbiol.* 61: 145–154.

17 Datta, N. and R.W. Hedges (1972): "Host ranges of R factors," *J. Gen. Microbiol.* 70: 453–460.

18 Deley, J. (1968): "Molecular biology and bacterial phylogeny," *Evolution Biol.* 2: 103–156.

19 Dose, K.S., W. Fot, G.A. Deborin, and T.T. Pavlovskaya (1974): *The Origin of Life and Evolutionary Biochemistry.* New York: Plenum Pub. Corp.

20 Falkow, S., *et al.* (1977): "The transposition of ampicillin resistance: Nature of ampicillin resistant *Harmophilus influenzae* and *Neisseria gonorrhoeae,*" *Topics of Infectious Diseases* 2: 115–129.

21 Freeman, V.J. (1951): "Studies on virulence of bacteriophage infected strains of *Corynebacterium diphtheriae,*" *J. Bacteriol.* 61: 675–688.

22 Goret, P. and L. Joubert (1949): "Le sol, être vivant méconnu," *Rev. Med. Vet.* 100: 594–595.

23 Gunsalus, I.C., M. Hermann, W.S. Toscano, Jr., D. Katz, and G.K. Gary (1975): "Plasmids and metabolic diversity," in D. Schlessinger, ed., *Microbiology—1974.* Washington, D.C.: Amer. Soc. for Microbiol., pp. 207–212.

24 Hungate, R.E. (1966): *Rumen and its Microbes.* New York: Academic Press.

25 Jacob, R. and J. Monod (1961): "Genetic regulatory mechanism in the synthesis of proteins," *J. Mol. Biol.* 3: 318–324.

26 Jones, D. and P.H.A. Sneath (1970): "Genetic transfer and bacterial taxonomy," *Bacteriol. Rev.* 34: 40–81.

27 Karska-Wysocki, B. and S. Sonea (1973): "Sensibilité à l'action létale des rayons UV d'une souche de *Staphylococcus aureus* en relation avec le nombre des propages présents," *Rev. Can. Biol.* 32: 151–156.

28 Lederberg, J. and E.M. Tatum (1946): "Gene recombination in *E. coli*," *Nature* 158: 558.

29 Ledoux, L. (1971): *Informative Molecules in Biological Systems*. New York: Elsevier-North Holland Publishing Co.

30 Lovelock, J.E. and Margulis, L. (1974): "Atmospheric homeostasis by and for the biosphere: the Gaia hypothesis," *Tellus* 26: 2–10.

31 Lwoff, A. (1953): "Lysogeny," *Bacteriol. Rev.* 17: 209–337.

32 Margulis, L. (1970): *Origin of Eukaryotic Cells*. New Haven, Conn.: Yale University Press.

33 Margulis, L. (1974): "The classification and evolution of prokaryotes and eukaryotes," in R.C. King, ed., *Handbook of Genetics*, Vol. 1, Bacteria, bacteriophages and viruses. New York, N.Y.: Plenum Press, pp. 1–41.

34 Margulis, L. (1975): "The microbes' contribution to evolution," *Bio. Systems* 7: 266–292.

35 Margulis, L. (1981): *Symbiosis in Cell Evolution*. San Francisco: W.H. Freeman and Co.

36 Margulis, L. (1982): *Early Life*. Boston: Science Books International.

37 Mourant, A.E. (1971): "Transduction and skeletal evolution," *Nature* 231: 466–467.

38 Muller, J.H. (1964): "The relation of recombination to mutational advance," *Mutat. Res.* 1: 2–9.

39 Oparin, A.I., A.G. Pasynskil, A.E. Braunshtein, T.E. Pavlovskaya, F. Clark, and K.L.M. Sunge (1959): *The Origin of Life on Earth*. New York: Macmillan.

40 Reanney, D.C. (1974): "On the origin of prokaryotes," *J. Theor. Biol.* 48: 243–251.

41 Reanney, D.C. (1974): "Viruses and evolution," *Rev. Cytol.* 37: 21–43.

42 Reanney, D.C. (1976): "Extrachromosomal elements as possible agents of adaptation and development," *Bact. Rev.* 40: 552–590.

43 Richmond, M.H. (1970): "Plasmids and chromosomes in prokaryotic cells," in Charles, H.P. and B.C.J.G. Knight, ed., *Organization and Control in Prokaryotic and Eukaryotic Cells*. Cambridge: Cambridge University Press.

44 Richmond, M.H. and B. Wiedeman (1974): "Plasmids and bacterial evolution," *Symp. Soc. Gen. Microbiol.* 24: 59–85.

45 Sneath, P.H.A. (1957): "Some thoughts on bacterial classification," *J. Gen. Microbiol.* 17: 184–200.

46 Sneath, P.H.A. (1974): "Phylogeny of microorganisms," *Symp. Soc. Gen. Microbiol.* 24: 1–39.

47 Sonea, S. (1971): "A tentative unifying view of bacteria," *Rev. Can. Biol.* 30: 239–244.

48 Sonea, S. (1972): "Bacterial plasmids instrumental in the origin of eukaryotes," *Rev. Can. Biol.* 31: 61–63.

49 Sonea, S., M. Desrochers, and B. Karska-Wysocki (1974): "Augmentation de l'effet bactéricide de la chaleur chez les souches polylysogènes de *Staphylococcus aureus* et *Bacterium anitratum*," *Rev. Can. Biol.* 33: 81–85.

50 Sonea, S. and M. Panisset (1976): "Manifesto for a new bacteriology," *Rev. Can. Biol.* 35: 158–167.

51 Sonea, S. and M. Panisset (1980): *Introduction à la nouvelle bactériologie.* Montreal: Presses de l'Université de Montréal, and Paris: Masson.

52 Stanier, R.Y. (1971): "Towards an evolutionary taxonomy of the bacteria," in A. Pérez-Miravette and D. Pelzez, ed., *Recent Advances in Microbiology.* Mexico: Ass. Mexicana de Microbiol.

53 Stanier, R.Y. (1980): "The journey, not the arrival, matters," *Ann. Rev. Microbiol.* 34: 1–48.

54 Stanier, R.Y., E.A. Adelberg, and J.L. Ingraham (1976): *The Microbial World.* Englewood Cliffs, N.J.: Prentice-Hall, Inc.

55 Taylor, F.J.R. (1974): "Implication and extensions of the serial endosymbiosis theory of the origin of eukaryotes," *Taxon* 23: 229–258.

56 Walcott, C.D. (1914): "Pre-Cambrian Algonkian algal flora," *Smithsonian, Misc. Coll.* 2: 64–68.

57 Watanabe, T. (1963): "Infective heredity of multiple drug resistance in bacteria," *Bacteriol. Rev.* 27: 87–115.

58 Whittaker, R.H. and L. Margulis (1978): "Protist classification and kingdoms of organisms," *Bio. Systems* 10: 3–18.

59 Woese, C.R., L.C. Magrum, and G.E. Fox (1978): "Archaebacteria," *J. Molecular Evolution* 11: 245–252.

Index